The Executive Guide to Healthcare Kaizen

Leadership for a Continuously Learning and Improving Organization

The Executive Guide to Healthcare Kaizen

Mark Graban and Joseph Swartz

"I believe that Kaizen is essentially a "human business." Management must meet diversified requirements of its employees, customers, stakeholders, suppliers, and its community. In this sense, the healthcare profession can probably best benefit from Kaizen because its central task is people. I am honored to write the foreword to *Healthcare Kaizen* by Mark Graban and Joseph Swartz."

Masaaki Imai
Author of *KAIZEN* and *Gemba Kaizen*

"To get started with Kaizen, you should do the following. First, read this book. Second, ask your employees to read the book. Third, ask your employees to begin a Kaizen system. It is just that simple. You just ask, and you will get what you ask for. Just do it and learn from the process."

Norman Bodek
Author of *How to do Kaizen* and *The Harada Method*

"I hope you will discover, as we have, the incredible creativity that can be derived by engaging and supporting each and every employee in improvements that they themselves lead."

Robert J. (Bob) Brody
CEO, Franciscan St. Francis Health

"At a time when many hospitals and health systems have relegated Lean to the "Project of the Month Club", Graban and Swartz remind us of the fundamentals that help organizations keep their Lean initiatives alive and thriving. I hope everyone reads this book and recommits to the fundamentals of Lean, particularly the involvement of frontline staff in process redesign."

Fred Slunecka
Chief Operating Officer, Avera Health

"Unleashing the energy and creativity of every employee to solve problems everyday should be the sole focus of every healthcare leader. Unfortunately, there are only a handful of examples where this is happening. *Healthcare Kaizen* provides examples of front line staff coming up with solutions to problems on their own and implementing them. Healthcare leaders need to read this book to understand that their management role must radically change to one of supporting daily kaizen if quality safety and cost are to improve in healthcare."

John Toussaint, MD
CEO, ThedaCare Center for Healthcare Value
Author, *On the Mend* and *Potent Medicine*

"In *Healthcare Kaizen*, Mark Graban and Joseph Swartz show us that Kaizen is more than a set of tools. What we have learned through our application of the Virginia Mason Production System is that Kaizen is a management methodology of continuous improvement that must permeate the fabric of the entire organization. Front line staff must know, understand, embrace and drive Kaizen and its tools to achieve incremental and continuous improvements. This book will help health care organizations around the world begin and advance their journey."

Gary Kaplan, MD, FACP, FACMPE, FACPE
Chairman and CEO, Virginia Mason Medical Center

"The healthcare industry is in the midst of truly fundamental change, and those organizations that engage their front line staff in developing the strategies for improving care, enhancing satisfaction, and streamlining processes to reduce unnecessary variation and expense will be well positioned to thrive in a post-reform environment. In their book, *Healthcare Kaizen*, Graban and Swartz create a roadmap for using incremental, staff driven changes to inculcate performance improvement into the culture of an organization in a sustainable manner. This book represents a wonderful resource for healthcare leaders looking to foster innovation at all levels."

Brett D. Lee, PhD, FACHE
CEO, Lake Pointe Health Network

"*Healthcare Kaizen* is a practical guide for senior healthcare leaders aspiring to engage frontline staff in true continuous improvement. Graban and Swartz skillfully illustrate how to foster and support daily continuous improvement in health care settings. Health systems struggle to move beyond improvement work being extra work done in "special projects" facilitated by experts. This book can guide organizational transformation so that continuous improvement becomes part of the daily work of frontline staff."

John E. Billi, MD
Associate Vice President for Medical Affairs
University of Michigan

"When healthcare organizations take initial steps on their Lean journey, they often focus very heavily on tools and grand solutions, which may create new barriers to innovation. In *Healthcare Kaizen*, Mark and Joe remind us of the great power of daily problem solving. Their examples reinforce that learning is a result of the repeated tests of changes that are often small and simple, and less often by hitting the home runs of improvement. The story of Franciscan St. Francis Health is compelling, where leaders created the opportunity for great people at the frontline making great improvements for patient care."

Michel Tétreault, MD, President and CEO
Bruce Roe, MD, Chief Medical Officer
St. Boniface Hospital, Winnipeg, Canada

"Without exception, the leadership of the health system is the determinant of success or failure in Lean transformation. *The Executive Guide to Healthcare Kaizen* is a focused and concise guide for that journey, a must read for those who have that responsibility."

Dave Munch, MD
Senior Vice President and Chief Clinical Officer
Healthcare Performance Partners

"In the last decade, implementation of the Lean production model in a healthcare setting has produced remarkable outcomes and revolutionized the way we deliver care. Using examples from Franciscan Health and other forward-thinking medical groups, the book contains valuable strategies for organization-wide cultural transformation to create a more efficient, patient-centered healthcare system dedicated to continuous quality improvement."

Donald W. Fisher, PhD
President and CEO
American Medical Group Association

"Mark Graban and Joseph Swartz have brought to life the critical concept of kaizen – continuous improvement. In this latest edition, a great deal of emphasis is placed on senior management engagement and support of ongoing improvement. Most agree that meaningful, sustained change cannot occur without leadership from the top, engagement of the front lines, and cohesion of the leadership chain. This book does a wonderful job of delivering these important concepts in an accessible, intriguing manner. Kudos to Graban and Swartz!"

Jody Crane, MD, MBA
Senior Medical Director, Stafford Hospital
Principal, X32 Healthcare
Co-author, *The Definitive Guide to Emergency Department Operational Improvement*

"Unfortunately the lean movement has too often turned into a race to implement as many of the tools of lean in as many places as possible. This is totally alien to the spirit of kaizen or the purpose of the Toyota Production System. The purpose is to create a culture of continuous improvement with people at all levels thinking deeply about their ideal vision for the people and process, and purposefully taking steps to achieve the vision. The vision should be for the good of the enterprise, not to check the box for the lean folks who are auditing 5S and visual management.

Mark and Joe have a deep understanding of the purpose of TPS and what is needed in healthcare to raise this from a program to a true culture that can tackle all the difficult challenges that face modern medicine. He has been steeped in the healthcare field for years and has great examples to illustrate kaizen, both small and big changes. In this book he takes on the challenge of driving kaizen down to the level of every work group--truly the deepest meaning of kaizen. This takes exceptional leadership, a second nature understanding of the tools, and always working at the gemba to solve the real problems. Hopefully this book will become a blueprint for healthcare organizations everywhere that truly want to be great!"

Jeffrey Liker
Professor of Industrial and Operations Engineering
University of Michigan
Author of *The Toyota Way*

"Adoption of the Lean philosophy is dead on arrival without the involvement of an organization's senior leadership. Yet, what are members of the executive suite to think when a bunch of Japanese terms coming flying past their desk? And when the leadership philosophy required is something quite different from their training and experience? Graban and Swartz help cut through all this in a presentation that is cogent, efficient, and thoughtful. Whether you are new to Lean principles or experienced in them, this book has something to offer. Even if you don't choose to take the entire Lean journey, you will receive insights and ideas that will help you get better results from your organization."

Paul F. Levy
Author of *Goal Play! Leadership Lessons from the Soccer Field*
Former CEO, Beth Israel Deaconess Medical Center

"It has been studied and shown that true north for healthcare organizations is an engaged senior leader and senior leadership team. This factor alone is the difference between mediocrity and excellence when it comes to performance and sustained extraordinary metrics for care, health and cost. *The Executive Guide to Healthcare Kaizen* provides a foundation for you as an executive to build the learning organization needed in today's environment. Smart, to the point, and handy. You will find this guide invaluable."

Betty Brown MBA MSN RN CPHQ FNAHQ
Immediate Past President NAHQ
Principal, ELLO Consulting, LLC

"At Beth Israel Deaconess Medical Center, everybody improving every day is a critical aspect of our Lean and quality improvement efforts. *Healthcare Kaizen*, is full of relatable examples as well as practical ideas that will inspire staff, clinicians and leaders at all levels. Its' must-have supplement, *The Executive Guide to Healthcare Kaizen: Leadership for a Continuously Learning and Improving Organization*, clearly outlines the role of management in leading this important work. It is not enough to be supportive; rather, one must demonstrate genuine interest with active participation and not delegate continuous improvement to others."

Alice Lee
Vice President, Business Transformation
Beth Israel Deaconess Medical Center

"For the past 7 years I have been leading a successful lean healthcare transformation at Chugachmiut, the non-profit organization I lead in Alaska. During that time, I have learned that respect for the people who work for you is key to any transformation. Mark Graban and Joseph Swartz do a great job of capturing this truth in their book, *Healthcare Kaizen: Engaging Front-Line Staff in Sustainable Continuous Improvements*. Every employee can learn the tools of lean, and improve processes as a result. However, sustaining a lean transformation and resisting entropy requires engaging front line employees in a long term vision for serving their customers and in true continuous improvement. Employees who work in a culture that removes blame and shame, operates on facts and seeks improvement continuously have great leadership and will respond with incredible results. This book is a long needed addition to my growing lean healthcare library."

Patrick M. Anderson
"Lean in Alaska"
Governance and Management Consulting

"The term 'kaizen' has been interpreted in many ways since we learned of the Toyota Production System in healthcare. Mark and Joe demystify the term, help us understand its real meaning, and help us see how using kaizen can help us improve in healthcare and, frankly, how we can use kaizen to save lives. The philosophy, tools and techniques discussed in the book work, and work well, in any environment. We in healthcare must improve - we owe it to our patients and communities - and Mark and Joe are helping to show us the way."

Dean Bliss
Chief Quality Officer
Four Oaks

"The healthcare industry has long struggled to tap one of the biggest sources available to it for ideas to improve outcomes and reduce costs – its front-line staff. Healthcare Kaizen lays out a step-by-step approach that any healthcare organization can use to get the dramatic results that come when its workforce is fully engaged in kaizen activities on a daily basis. This inspirational book is packed with examples and is informed by the authors' years of experience on the "front-lines" themselves, helping leading healthcare organizations around the world to build successful kaizen programs."

Alan G. Robinson, PhD
Professor, Isenberg School of Management at the University of Massachusetts
Co-author, *Ideas Are Free* and *Corporate Creativity*

"What Mark Graban and Joseph Swartz have done in *Healthcare Kaizen* and *The Executive Guide to Healthcare Kaizen* is to bring hope and light to a part of our society that is facing increasing challenges. Full of examples and illustrations from hospitals and healthcare professionals leading the way in the journey to patient-centered, error-free care delivery, this book makes it easy to connect with this very powerful concept of kaizen. By putting kaizen within the broader tradition of quality improvement, shedding light on its historical development and pointing out potential pitfalls in its application in healthcare, the authors provide a great service to the healthcare community. I was especially impressed by the authors' important insights on what a kaizen culture feels like, and how people at all levels can and must engage in daily improvement. These books will be a reference on the subject for many years to come."

Jon Miller
CEO, *Kaizen Institute*

"At last, a crystal clear description of Kaizen as a philosophy and a work culture, not another top-down tool. Graban and Swartz show, in unequivocal detail, that Kaizen need not be viewed as a formal, five-day event, requiring X, Y, and Z participants, components, and steps. The compelling examples from Franciscan Health and others paint a picture of a hospital culture steeped in respect for people and continuous improvement—the very elements of Lean, Kaizen, and scientific inquiry. By busting the myth of the five-day "event," the authors show the true, sweeping potential of Kaizen in the healthcare workplace."

Naida Grunden
Author, *The Pittsburgh Way to Efficient Healthcare*
Co-author, *Lean-Led Hospital Design: Creating the Efficient Hospital of the Future*

"The vision of a world in which our healthcare institutions operate with a universal discipline of relentless, patient-centered improvement remains a vitally important yet distant dream. In *Healthcare Kaizen*, Mark Graban and Joseph E. Swartz illustrate just how to make that dream a reality."

Matthew E. May
Author of *The Elegant Solution* and *The Laws of Subtraction*

When embarking on a Lean transformation, one of the greatest leadership and cultural challenges is getting to the point where the frontlines have the skills for and are truly authorized to make daily improvement. This shift not only accelerates results, it deeply engages the workforce, a precondition for achieving organizational excellence. Graban and Swartz present the kaizen philosophy in the most accessible way I've seen yet. They present a powerful model for preparing managers for their new role as improvement coaches and the frontlines for taking a far more active role in delivering greater value to the healthcare industry's various customers. THIS is the missing link in healthcare reform.

Karen Martin
Author,*The Outstanding Organization* and
The Kaizen Event Planner

The Executive Guide to Healthcare Kaizen

Leadership for a Continuously Learning and Improving Organization

Mark Graban and Joseph E. Swartz

Introduction by
Gary M. Kaplan, MD, Chairman and CEO,
Virginia Mason Medical Center, Seattle

CRC Press is an imprint of the
Taylor & Francis Group, an **informa** business
A PRODUCTIVITY PRESS BOOK

CRC Press
Taylor & Francis Group
6000 Broken Sound Parkway NW, Suite 300
Boca Raton, FL 33487-2742

Printed on acid-free paper
Version Date: 20130709

International Standard Book Number-13: 978-1-4665-8641-3 (Paperback)

Library of Congress Cataloging-in-Publication Data

Graban, Mark, author.
The executive guide to healthcare kaizen : leadership for a continuously learning and improving organization / Mark Graban, Joseph E. Swartz.
p. ; cm.
Includes bibliographical references and index.
ISBN 978-1-4665-8641-3 (hardcover : alk. paper)
I. Swartz, Joseph E., author. II. Title.
[DNLM: 1. Total Quality Management. 2. Hospital Administration. 3. Leadership. WX 153]

RA971
362.11068--dc23 2013026192

Visit the Taylor & Francis Web site at
http://www.taylorandfrancis.com

and the CRC Press Web site at
http://www.crcpress.com

Contents

Foreword

In my book *KAIZEN, The Key to Japan's Competitive Success* (McGraw-Hill, 1986), I ended with the following words

> It is my sincere hope that we will be able to overcome our "primitive" state and that the Kaizen strategy will eventually find application not only in the business community, but also in all institutions and societies all over the world.

Look over the last 25 years since its publication; I am profoundly frustrated with the slow pace at which Kaizen strategy has been embraced by the business community. On the other hand, I am encouraged to note that Kaizen is rapidly gaining momentum in nonbusiness institutions like healthcare, services, and government.

I believe that Kaizen is essentially a *human business*. Management must meet diversified requirements of its employees, customers, stakeholders, suppliers, and its community. In this sense, the healthcare profession can probably best benefit from Kaizen since its central task is people. I am honored to write a foreword to this book by Mark Graban and Joseph Swartz.

Taking this opportunity, I wish to mention a few reminders to successfully embrace the Kaizen strategy.

1. Embracing Kaizen is a long-term journey. It is not a flavor of the month and requires the cultural change, commitment, and self-discipline that needs to be sustained over many decades until they become routine business practices.
2. Top management commitment is the only way to successfully embrace Kaizen, without which nothing else you do will matter.
3. We need to approach our daily business in two phases: one is to maintain the status quo, in which the standard (the best way to do the job) is established and followed. This process is called maintenance and requires dedicated management effort to sustain it, but it is often overlooked or belittled.

The second phase is Kaizen, which means to find a better way and revise the current standard. Thus, to maintain and improve the standard becomes the main task of management.

4. My definition of Lean is to employ minimum resources for the maximum benefits. Therefore, Kaizen leads to Lean and Lean leads to green. Kaizen is the most environmentally friendly approach.
5. Welcome problems. The more problems, the better, since we have more Kaizen opportunities. We only need to establish priorities in dealing with problems. When the problem is correctly identified, the project is halfway successful.
6. One of the best ways to identify problems is to observe the flow of operations. In medical institutions, there are many types of flows such as information, physical movement of patients and families, medicines, and supplies. Wherever and whenever the flow is disrupted, there is a Kaizen opportunity.

   A majority of disruptions of the flow can be easily detected and solved with common sense and do not require sophisticated technologies.
7. Remove the barriers between professionals and laymen.

I sincerely hope that you will find your Kaizen journey to be challenging, but most rewarding.

Masaaki Imai
Chairman, Kaizen Institute

Preface

A *culture of continuous improvement* is something that many healthcare organizations have talked about, as evidenced by statements that can be found on hospital websites and lobby signs. Creating this culture is easier said than done, but turning that nice-sounding ideal into reality is becoming increasingly urgent in healthcare.

This book is for every leader who wants to create a continuously learning and improving organization that will not just survive but also thrive in an increasingly challenging and uncertain future. We agree with the Institute of Medicine that, "Americans would be better served by a more nimble health care system that is consistently reliable and that constantly, systematically, and seamlessly improves."[1] We think that statement also applies around the world, as do the methods in this book that will help get you there.

Healthcare organizations face many threats, so they need many types of improvement—large strategic initiatives, formal team-based projects, and smaller local improvements. All three types of improvement are complementary and necessary to provide the best value to patients, the best workplaces for staff, and the strongest competitive positions for our organizations.

Across industries, many organizations that faced increasingly competitive markets have turned to Lean, Kaizen, and other process improvement practices in order to both radically rethink and continuously improve the way they run their businesses—to improve both quality and cost. Now, the same success stories are happening in healthcare, or least in some pockets around the world, as coauthor Mark wrote about in *Lean Hospitals*.[2] The problem is that many (if not most) of these Lean initiatives are completely project based, missing the opportunity for truly continuous improvement that engages each employee and leader—*Kaizen*.

Keith Jewell, chief operating officer of Franciscan St. Francis Health System, says, "Healthcare reform is at our front doors and we, as executives, have to get everyone ready to move with us at an ever faster pace. Kaizen is not sufficient by itself, but by getting the workforce engaged in small change leads to them being more involved, as well as becoming more accepting of (and more of a part of) the major changes that are coming. Kaizen is an important part of building a culture that is required to accept change and transformation."[3]

Repeating Patterns from Other Industries

Today, American healthcare organizations are facing many newly aggressive competitors that are putting pressure on the bottom line. A very similar story happened to Xerox 30 years ago at the hands of an emergent competitor, Canon, and to Harley Davidson at the hands of Yamaha and Suzuki. At that same time, the story repeated itself in the American auto industry as Toyota and Honda grew dominant. Ten years later, it was happening to every American company trying to compete with those that had shifted manufacturing to China.

The challenge facing the American auto industry was not just about cost, it was also about quality. Many people swore off buying American cars after bad experiences in the 1970s and 1980s. Japanese car companies took advantage of this opening in the market to deliver high-quality cars that also cost less.[4] Many of the Detroit automotive executives thought it just wasn't possible to have better cars that cost less, so they concluded, incorrectly, that the Japanese were being predatory by selling below cost. Out of frustration, some Americans actually smashed Japanese cars with sledgehammers in front of the U.S. Capitol. While American automakers have made great strides in quality in recent years, they lost at least a decade spent making excuses and blaming others instead of doing everything they could to put their own house in order.

The Need to Bend Cost *and* Quality Curves

An executive at a Boston hospital recently lamented to coauthor Joe that they had lost up to 15% of their business in key areas over the past year and they were continuing to lose market share. Newly competitive hospitals in the suburbs were taking volume away by offering prices as much as 29% lower.[5] Massachusetts has been on the forefront of healthcare reform for many years. ThedaCare, a health system in Wisconsin, reduced costs for coronary bypass surgery by 22% while also reducing mortality from 4% to zero. They also reduced the cost of inpatient care by 30% while improving quality and patient satisfaction.[6]

Patients today are becoming more aware of the preventable harm that occurs far too often in modern hospitals.[7] The government and other payers are refusing to pay for a wide range of "never events," proven to be preventable with better processes and less dysfunctional organizational cultures. Consumers have had little information or data available that is useful in choosing a healthcare facility or a surgeon. The trend toward more transparency, as seen, for example, in Wisconsin,[8] creates the threat that organizations with poor, or slowly improving, quality will be punished in the reformed healthcare market. Reform is changing the rules of the game, shifting the healthcare market from a volume-based model to a value-based model.

Hospitals also face the growing threat of patients being given lucrative incentives to travel to a location where healthcare is arguably better and definitely less expensive—this could be overseas in India or up the road within Wisconsin![9] Walmart, which has seen "huge variation" in healthcare quality and cost, struck a deal to fly employees to Seattle for specialty care at Virginia Mason Medical Center (led by Gary Kaplan, MD, who wrote the Introduction for this book), one of the world's leaders in applying Lean and Kaizen, as one of their six centers of excellence organizations that provide "good value in terms of quality and cost of care."[10]

Price and quality will matter more as healthcare reform moves along and customers become aware of increasing gaps in quality performance between average hospitals and those that have reduced "door to balloon" times by 60% or have reduced central line infections by 95%. Healthcare exchanges are coming, in which insurance plans will be selected by consumers based on side-by-side comparisons from a wide variety of insurance companies and hospital systems. Who will those customers select? What healthcare providers will be included in the increasingly competitive insurance plans? Organizations can make excuses or they can take action—engaging all of their leaders, staff, and physicians in the process.

Are You Ready to Compete?

Will you be ready to respond if a competing healthcare system is offering a comprehensive healthcare plan that is priced at up to 29% less than yours? Will you be ready if an employer wants to send its patients to another city where both the infection rates and cost for a joint replacement procedure are lower? Are you ready for whatever unforeseen threat appears a few years from now?

Companies like Xerox, Harley Davidson, and Ford stepped up to their competitive threats and survived, but thousands of other organizations did not have enough cash reserves, enough customer goodwill, or enough resolve to last long enough to fundamentally redesign their businesses to be competitive. Hospitals are at a point where they can't afford to wait and see. Overhead costs are high, margins are fragilely low, quality needs to improve, and change is coming faster than most organizations are designed to handle. Barbarians will soon be at the gates of organizations that can't change fast enough. The era of being able to sit back and react to the details in the regulations that end up finding their way through the legislative maze is going away and the survivors will be those who proactively and aggressively learn to improve everything they do at a faster and faster rate.

Rachelle Schultz, CEO at Winona Health (Minnesota), says, "Lean and Kaizen is a solid and powerful approach to dealing with the magnitude of change coming and the lack of details around it. We will be far better prepared for breakthrough thinking and change."[11]

What's Needed: Leadership and Transformation

To change at the rate required to survive, we contend that every employee in a healthcare organization will need to be engaged at his or her highest level of contribution to rethink and redesign everything the organization does. That is what Canon, Yamaha, and Toyota did, and that is what Xerox, Harley Davidson, and Ford learned to do. These companies didn't just overhaul their factories, they also updated their products, changed their cultures, and overhauled their management systems and leadership styles.

This book is about putting the infrastructure and practices in place to develop an organizational culture that continually and forever engages all employees to improve their work and patient care. It's about the mechanisms and, more importantly, the leadership behaviors required for transformation. Creating this culture takes years. Do you have time to wait?

Here are some questions to discuss with your board and your team, courtesy of David Munch, MD, former chief clinical and quality officer at Exempla Lutheran Medical Center:

1. Do you think fundamental changes need to occur in your organization to meet the challenges of healthcare reform and changing demographics?
2. When do you think these changes will be on your doorstep?
3. How much time do you think it will take to change your organization?
4. What are you waiting for?
5. What changes will you and your executive team need to make personally for this to be successful?
6. What about the leaders and managers who work for you?[12]

These are important questions and we hope this book helps you face these challenging times. We wish you great success in this important transformation—your patients, your staff and physicians, and the future of your organization are counting on you and your fellow leaders.

Endnotes

1 Institute of Medicine, "Report Brief: Best Care at Lower Cost: The Path to Continuously Learning Health Care in America," September 2012, http://www.iom.edu/~/media/Files/Report%20Files/2012/Best-Care/BestCareReportBrief.pdf (accessed January 5, 2013), 1.

2 Graban, Mark, *Lean Hospitals: Improving Quality, Patient Safety, and Employee Engagement* (New York: Productivity Press, 2011), 5.

3 Jewell, Keith, personal interview, December 2012.

4 Womack, James P., Daniel T. Jones, and Daniel Roos, *The Machine That Changed the World: The Story of Lean Production* (New York: Harper Perennial, 1991), 81.

5 Mohl, Bruce, "Cerberus's Health Care Play," *CommonWealth*, July 10, 2012, http://www.commonwealthmagazine.org/News-and-Features/Features/2012/Summer/001-Cerberuss-health-care-play.aspx (accessed January 18, 2013).

6 Toussaint, John, "Writing the New Playbook for U.S. Health Care: Lessons from Wisconsin," *Health Affairs* 28, no. 5 (September/October 2009): 1343–1350, doi: 10.1377/hlthaff.28.5.1343.

7 Graban, Mark, "A Lean Guy Watches 'CNN's 25 Shocking Medical Mistakes,'" http://www.leanblog.org/2012/06/cnns-25-shocking-medical-mistakes/ (accessed January 27, 2013).

8 Wisconsin Health Information Organization, http://www.wisconsinhealthinfo.org/ (accessed January 29, 2013).

9 Torinus, John, *The Company That Solved Health Care: How Serigraph Dramatically Reduced Skyrocketing Costs While Providing Better Care, and How Every Company Can Do the Same* (Dallas: BenBella Books, 2010), 44.

10 Ostrom, Carol M., "Virginia Mason among Top Hospitals Tapped by Walmart," http://seattletimes.com/html/localnews/2019403172_walhealth12m.html (accessed January 27, 2013).

11 Graban, Mark, "Podcast #164–Rachelle Schultz, CEO of Winona Health," LeanBlog.org, http://leanblog.org (accessed May 5, 2013).

12 Munch, David, personal interview, January 2013.

Acknowledgments from Mark Graban

There are many people I need to thank, first of all my coauthor Joe Swartz. This book has been a wonderful partnership since I first raised the idea at the Society for Health Systems annual conference in early 2009. Without Joe's passion and hard work for Kaizen at the Franciscan St. Francis Health System, this book would not have the depth of his years of experience and lessons learned.

I also want to thank my mentors in Kaizen, including Norman Bodek, whose never-ending enthusiasm and his belief in everybody's ability to participate in Kaizen have been inspiring. Thanks also go to Norman for his formal training and certification in "Quick and Easy Kaizen" and to Rick Malik, the worldwide director of ValuMetrix Services, for his support in allowing me to take Norman's workshop. Norman first introduced me to my coauthor, Joe, back in 2005 because Norman knew we were both transitioning from manufacturing into healthcare, so I appreciate his networking and sharing.

Many thanks are also due to my other mentors and role models in my study and practice of Kaizen, including John Shook and Jim Womack of the Lean Enterprise Institute, Pascal Dennis of Lean Pathways, Inc., and John Toussaint, MD, of the ThedaCare Center for Healthcare Value.

Thanks to all who graciously shared examples of Kaizen, either anonymously or with recognition for their organizations. Most of them are quoted or cited in the book. Without their passion for improving patient care and creating a better healthcare workplace, along with their willingness to share with others, this book would not be possible.

Thanks go also to my trusted colleagues and friends who reviewed early drafts and provided valuable feedback and advice on our earlier *Healthcare Kaizen* and this edition, including John Toussaint, MD, David Munch, MD, Paul Levy, Brian Buck, Lewis Lefteroff, Ken LePage, Jim Adams, Michael Lombard, Greg Jacobson, Karen Martin, Helen Zak, and Naida Grunden. I also want to thank Jon Miller, of the Kaizen Institute, for helping secure Masaaki Imai's support and foreword. Thanks to my editor Kris Mednansky for her support on this project and for my first book, *Lean Hospitals*.

Finally, I want to thank my wonderful and inspiring wife, Amy, for her love and support through this and all other writing projects, and for her patience about all of the time I was off in my office writing. Thanks also go to my in-laws, Charlie and Debbie, for their encouragement and support during the writing process. I also want to acknowledge my parents, Bob and Marlene, for their love and the educational opportunities they created for me.

I would like to dedicate this book to the caring, hardworking healthcare professionals around the world. Thank you for your service to your patients and communities. Thanks especially for your dedication to process improvement and for working toward providing the best patient care possible.

Acknowledgments from Joe Swartz

I am grateful for the opportunity to have worked with my coauthor, Mark, on this book. Although an extremely accomplished and highly intelligent person, he treated me graciously as a full partner throughout this work. I appreciate his patience, his extensive expertise, and his ability to listen to the many voices in the domain of Kaizen and to discern and continually pull together the vision for this book. He is an inspiration.

I bow to honor Franciscan St. Francis Health. To their leadership, Bob Brody, Keith Jewell, and Paul Strange, MD, for giving me the opportunity to participate in their amazing organization and mission. To the staff for significant and meaningful Kaizen contributions, which enrich each reader's ability to translate Kaizen to his or her unique workplace.

Many heartfelt thanks to my colleagues and friends—to my team at St. Francis for helping me develop the Kaizen program showcased in this book: Mischelle Frank, Julia Dearing, Tom Pearson, and Matthew Pierce. To Heather Woodward for introducing me to Franciscan. To Marcia Ellett, writer and wonderful friend, who encouraged me throughout the writing of this book. To Jim Huntzinger, Brian Hudson, and Tim Martin whose deep Lean knowledge helped me keep Kaizen real. To Dan Lafever for calling me sensei when I should be calling *him* sensei. To Norm Bodek who continually encouraged me to write. To John Feller, MD, and Chuck Dietzen, MD, childhood and best friends who have been by my side coaching me on the high-leverage knowledge of healthcare.

I wish to thank those who carved time out of their busy day to review selected chapters and gave such good feedback: Diana McClure, Laura Louis, Paula Stanfill, Joe Click, Nik Janek, Kyle Ellen Brown, and Amy Lynelle.

To my parents, James B. and Kathleen Swartz, who set the pattern for the person I am and who inspired me to write; to my children, Jordan, Paul, and Madison, who are the loves of my life; to my brother Greg and sisters Laura and Julie, who have been such a joy in my life; to my friends for simply being my friends; to God for life and possibilities.

Finally, I am appreciative for the privilege Mark and I had to learn the field of Lean and Kaizen by standing on the shoulders of giants and for having the opportunity to convey some of what we have learned from those giants.

Like Mark, I dedicate this book to the caring, hardworking healthcare professionals around the world. Thank you for continually sharing your life purpose and meaningful reasons why you are called into the field of healthcare. Thank you for your heart of service, care, and compassion. You have touched my family and me so much more significantly than you know.

About this Book

This book (*The Executive Guide to Healthcare Kaizen*) is a follow-up edition to our book *Healthcare Kaizen: Engaging Front-Line Staff in Sustainable Continuous Improvements*. This book, a derivative of that earlier work, is intended for senior leaders and other healthcare leaders who can initiate and sponsor a Kaizen program of staff engagement and continuous improvement for their organization.

Healthcare Kaizen is a longer book, with over 200 full-color illustrations, including 100 examples of real Kaizen improvements from different healthcare organizations. We, thankfully, received many positive reviews of the book, with the only real criticism being that the book was too complete or too heavy for an executive to carry onboard a flight.

This edition is intended to cover the "why" (why Kaizen?) and the "what" (what is Kaizen and a Kaizen culture?), while *Healthcare Kaizen* has more of the "how." There is significant overlap between this book and *Healthcare Kaizen*, so there is a consistent approach taught by both books. We envision *The Executive Guide to Healthcare Kaizen* as a smaller, more affordable, book that covers the basics of Kaizen, while *Healthcare Kaizen* is the more encyclopedic guide for practitioners.

Each chapter in this book starts with a "quick take" list of key points from the chapter. At the end of each chapter are discussion questions that we hope you will use to prompt discussion with your leadership team. We invite you to interact with us via our website, http://www.HCkaizen.com.

While this book has two coauthors, Mark Graban and Joe Swartz, we have done our best to write in a single voice. Any work that we directly participated in is written about in the third person to avoid confusion about who was writing particular sections.

Franciscan St. Francis Health is a three-hospital system in Indianapolis. Throughout this book, we will often refer to individual hospitals in this system in an abbreviated form like "Franciscan." That three-hospital system is part of the Franciscan Alliance, a 14-hospital system located throughout Indiana and northeastern Illinois. Where we focus on one of their other 11 hospitals, we will call them out uniquely, such as Franciscan St. Elizabeth Health.

Introduction

The value of using Kaizen to improve healthcare systems is indisputable. At Virginia Mason, we have been using Kaizen, based on the Toyota Production System, for more than a decade. The Virginia Mason Production System, as we call it, has allowed us to deliver safer, better, and more affordable care to our patients.

One of the keys to successful implementation of Kaizen is the serious commitment of leaders – including the CEO, senior executives, physician leaders, and boards of directors. In our organization, all leaders attend mandatory Kaizen training, are required to lead formal improvement events each year, and are expected to routinely coach and train employees about how to improve their work using Kaizen tools and methods. Kaizen is not a program or an activity that is the sole responsibility of one department; it is the management method used by all leaders at Virginia Mason to guide and operate every aspect of the organization.

Physician leadership is an important part of leadership commitment. An organization that reforms around physicians but doesn't involve them in the process will have difficulty succeeding in the long run. In my experience, the organizations with strong physician leadership and active physician involvement at all levels are best prepared to deliver change through Kaizen.

Kaizen tools encourage and guide change in day-to-day work by all employees. As employees gain a better understanding of Kaizen, they use its methods to improve how they do their work. At Virginia Mason, we encourage employees to record their improvement ideas as "Everyday Lean Ideas" and share them with the organization so they can be replicated across the medical center.

As it relates to employee engagement, Kaizen can't be imposed from above. Leaders should introduce, teach, and encourage the adoption of Kaizen methodologies, but it is only sustained when employees are engaged—because they have found that it makes their work easier and more satisfying. Employees become champions of Kaizen when they see it reduces the burden of work and the waste of rework and waste of time that come with inefficiency. It frees them to do the important things that add value for our patients and helps them recapture the passion that drove their original career decisions to work in healthcare.

Learning to use Kaizen consistently and effectively requires serious culture change and takes many years. This is really not unexpected, as using Kaizen requires deep organizational changes—changes that challenge long-held beliefs and many accepted practices. Our results have been gratifying and propel us to work even harder to deploy these methods and tools deeply within our organization. I believe this book will help any willing healthcare leader who sets out on the Kaizen journey to achieve similar success.

Gary S. Kaplan MD
Chairman and CEO, Virginia Mason Medical Center, Seattle

About the Authors

Mark Graban is an author, consultant, and speaker in the field of Lean healthcare. He is the author of *Lean Hospitals: Improving Quality, Patient Safety, and Employee Engagement* (2nd edition) and coauthor of *Healthcare Kaizen: Engaging Front-Line Staff in Sustainable Continuous Improvements*. Mark has worked as a consultant and coach to healthcare organizations throughout North America and Europe. He was formerly a senior fellow with the Lean Enterprise Institute and continues to serve as a faculty member. Mark is also the chief improvement officer for KaiNexus, a startup software company that helps healthcare organizations manage continuous improvement efforts. Mark earned a BS in industrial engineering from Northwestern University and an MS in mechanical engineering and an MBA from MIT Sloan Leaders for Global Operations Program. Visit his website at http://www.MarkGraban.com and his blog at http://www.LeanBlog.org.

Joseph E. Swartz is the director of Business Transformation for Franciscan St. Francis Health of Indianapolis, Indiana. He has been leading continuous improvement efforts for 18 years, including 7 years in healthcare, and has led more than 200 Lean and Six Sigma improvement projects. Joseph is the coauthor with Mark Graban of *Healthcare Kaizen: Engaging Front-Line Staff in Sustainable Continuous Improvement* and coauthor of *Seeing David in the Stone* and was previously an instructor at the University of Wisconsin. Joseph earned an MS in management from Purdue University as a Krannert Scholar for academic excellence.

Chapter 1

The Need for Kaizen

Quick Take

- Because healthcare faces such great challenges, we have no choice but to get everybody involved in identifying and implementing improvements.
- *Kaizen* is a Japanese word that means "change for the better."
- Kaizen is a key part of the Lean management philosophy and strategy.
- Kaizen engages all staff, physicians, and leaders in making improvements to safety, quality, access, and cost, while improving staff morale.
- Responsibility for Kaizen cannot be outsourced to consultants or delegated to a Quality Department.
- Kaizen creates higher engagement, which leads to better quality, which results in lower cost.
- The return on investment (ROI) on a Kaizen program can be impressive (millions per year with little investment).
- Kaizen creates a more flexible and adaptive organization, to better cope with conditions of extreme uncertainty.

> It is not the strongest of the species that survives, nor the most intelligent, but the one most responsive to change.
>
> **—Charles Darwin**

Franciscan St. Francis Health is a three-hospital system in Indianapolis, Indiana, that is part of the Franciscan Alliance, a 13-hospital system located throughout Indiana and northeastern Illinois. Founded over 135 years ago by a group of

Catholic Sisters, they take their inspiration from St. Francis of Assisi. Franciscan St. Francis Health has received numerous awards, including:

- Number one ranking in Indiana by HealthGrades for cardiac services and one of America's 100 best hospitals for cardiac care and coronary intervention.
- Top 5% in the nation and number one in Indiana for joint surgery, according to HealthGrades, 2007–2011.
- The HealthGrades Distinguished Hospital Award for Clinical Excellence™ in 2012 for being in the top 5% in the nation in overall clinical quality.

In 2005, after several years of suboptimal performance, the organization went looking for ways to capitalize on its foundation of excellence.

Paul Strange, MD, then Franciscan's vice president of quality, convinced the leadership team to launch a Lean Six Sigma program. Robert J. Brody, president and chief executive officer of St. Francis Health, and Keith Jewell, the chief operating officer, brought in a team of people from the outside, including one of this book's coauthors, Joe Swartz, and professors from Purdue University. Their Lean Six Sigma journey began in 2006, and Franciscan added a formal Kaizen program of continuous improvement in April 2007.

What started small, with a housekeeping staff member improving the way coffee filters were stored, has turned into an organization-wide program and, more importantly, a significant culture change. Today, Franciscan St. Francis empowers their staff members to identify problems and take action, leading to improvements both large and small. Some of these improvements made life a little better for the patients in a fun or charming way, some changed clinical practice, and others saved significant sums of money.

Since 2007, the three Franciscan St. Francis hospitals have implemented more than 17,000 improvements that have an estimated hard dollar cost savings of over $5.7 million, all with very little investment other than time, focus, and leadership. If every hospital in the United States could save $2 million a year through Kaizen, it would add up to $10 billion. That may seem like a small drop in the healthcare cost bucket, but the hard numbers from Franciscan don't include other "soft savings" such as staff time savings that can be reallocated to better patient care, increased patient and staff satisfaction, reduced error rates, and reduced waiting times.

The cost savings at Franciscan have been impressive, but cost is far from the primary goal of Franciscan's Kaizen culture. Throughout this book, we will be sharing stories from Franciscan and other hospitals to demonstrate how they engaged their staff in improvements that made a difference for all of their stakeholders: patients, employees, physicians, and the organization itself. CEO Bob Brody said, "There is every reason for any organization to encourage and support the Kaizen concepts. It creates a more efficient and productive work environment, a more satisfied patient, and a more satisfied workforce. These are linked to one another, and to hospital performance."[1]

For Franciscan and other leading organizations, Kaizen is just the right thing to do in terms of furthering its mission and fulfilling its desire to treat all staff members, physicians, and other stakeholders with the utmost of respect.

Kaizen = Change for the Better

The word *Kaizen* is translated from Japanese in a number of ways, most simply as "change for the better."[2] The Japanese characters are shown in Figure 1.1.

> Breaking down the word: *Kai* means "change," *zen* means "good."

A Kaizen is an improvement that is made by those who do the work. It is typically a small, low-cost, low-risk improvement that can be easily implemented.[3] Kaizen is an ongoing methodology and philosophy for challenging and empowering everyone in the organization to use their creative ideas to improve their daily work.

The word *Kaizen*, the way it is typically used, is synonymous with the phrase *continuous improvement*. An effective Kaizen approach is connected to measurable results and a deeper purpose. Children's Medical Center (Dallas, Texas) has a process improvement campaign that asks the simple question, "Is there a better way?" Clay York, manager of the core laboratory, and other leaders help tie the department's local improvement efforts to the organization's mission and purpose by asking team members if proposed changes will help provide "better care for kids."[4]

Figure 1.1 The word *Kaizen* in Japanese kanji characters.

Beyond the measurable results, Kaizen organizations value the organizational learning that results from the improvement process, as well as the personal learning and satisfaction of all who are involved.

IU Health Goshen Hospital has saved more than $30 million since 1998 with a program called "The Uncommon Leader" as part of its broader improvement program. In 2009, CEO James Dague, now retired, promised to shave his head if employees generated ideas that saved $3.5 million that year. The hospital more than doubled that savings goal, so Dague made good on that promise in front of his colleagues. The culture at Goshen has shifted to one where every person is empowered to make improvements to his or her daily work, making suggestions that can impact cost, quality, and patient care. For example, an emergency nurse educator saved $4,000 by changing the type of napkins used on patient trays and the GI department saved $22,000 by switching from disposable paper gowns to cloth gowns.[5] Goshen has gone 17 years without layoffs, undoubtedly being a key reason its employees are so enthusiastic about improvement.[6] Goshen was also named one of the top ten large employer workplaces in Indiana, in part due to its "workplace culture where employees feel valued."[7]

Kaizen = Meaningful Improvements

Paula Stanfill's husband had open-heart surgery. Paula is the manager of the neonatal intensive care unit (NICU) at Franciscan St. Francis Health. In the recovery room, Paula's husband could not speak because he was intubated with a breathing tube. He was trying to communicate by furrowing his eyebrows and squinting. He knew sign language and was motioning at his arm and trying to use his fingers to tell Paula something, but he could not make his hands do what he wanted them to do. Paula remembers her panic in realizing that something serious might be happening to her husband. He also began to panic, thinking the surgery had caused a serious problem with his arms. They were deeply distressed until the anesthesia wore off and he could speak again.

Paula learned that her husband's arms and hands were numb. He was a big man, and when the surgeons performed the procedure, they had leaned over his arms and put pressure on them, reducing the blood supply and causing the numbness. His arms remained numb for several weeks.

After this episode, Jessica Clendenen, a nurse in Franciscan's cardiac operating room, learned that several other patients had experienced similar postoperative numbness. She decided to do something about it. In January 2011, Jessica found some sled positioners that could be used to help tuck the patient's arms in place in a way that allowed the IV lines to be seen through the clear material. The use of positioners reduced the pressure on the patient, which meant improved quality, patient safety, and satisfaction.

This small, simple improvement can be described as a Kaizen. It was an improvement that made a difference to open-heart patients at Franciscan and was one that Paula will never forget. When she started making workplace improvements, Paula never realized it would touch her so personally. But after her husband's experience, Paula realized that the heart of Kaizen is the difference it can make in people's lives. Kaizen was no longer just a concept or a program to her; it had become a way of life.

Healthcare's Opportunity for Improvement

The largest room in the world is the room for improvement.

—Author unknown

As summarized in the Institute of Medicine's 2012 report titled, *Best Care at Lower Cost: The Path to Continuously Learning Health Care in America*, healthcare in the United States is underperforming in many ways, with problems including:

- $750 billion in "unnecessary health spending" in 2009
- 75,000 "needless deaths" that "could have been averted" in 2005 if every state performed as well as the best state[8]

Of the $750 billion in waste, $130 billion is estimated to come from "inefficiently delivered services," which includes mistakes, errors, and preventable complications, fragmented care, and operational inefficiencies.[9] These costs are in the span of control of health systems and represent a great opportunity for improvement.

While healthcare spending in the United States is far higher than any other country, rising costs, along with costs that are too high for national budgets, are a problem throughout the developed world. Diabetes costs alone "threaten to bankrupt" the National Health Service (NHS) in England[10] and the NHS is being forced to cut costs by £20 billion by 2015.[11] The budget cuts are leading to nurse layoffs and warnings from some NHS finance staff that patient care "will suffer" as a result.[12] Hospitals in Canada face government budget cuts, as public healthcare costs could comprise 70% of the Ontario provincial budget by 2022.[13] Michel Tétreault, CEO of St. Boniface General Hospital (Winnipeg, Manitoba), says his hospital has to, each year, deliver care to 4% more patients who are 4% sicker with the same staffing and resources.[14]

With American hospitals now facing the additional financial pressures of the Patient Protection and Affordable Care Act of 2010, also known as ObamaCare, reimbursements to health systems and physicians are being reduced, while penalties for preventable medical errors and readmissions are increasing.

The Institute of Medicine (IOM) report concluded that, "Left unchanged, health care will continue to underperform; cause unnecessary harm; and strain national, state, and family budgets."[15] Patients and their families are a powerful motivation for improving quality and safety, reducing costs, and improving access. Beyond the financial strain, there is also a big need to reduce the workplace strain and stress on physicians, nurses, and staff members in all departments. This stress and frustration is caused primarily by process issues that can be fixed by healthcare organizations—and can often be fixed by local teams, if they have a method for improvement and a conducive environment.

Today's health system executives, around the world, face the challenge of simultaneously improving across many dimensions. The old trade-offs can no longer hold true—the notion that better quality must inherently cost more or that the only way to increase capacity is spending more money on people, resources, and facilities. Leaders must figure out how to bust these old trade-offs and this requires new approaches. When facing large challenges, the most effective approach might, ironically, begin with the smallest of steps.

> Small improvements are believable and therefore achievable.[16]
>
> **—Tony Robbins**

The IOM's Recommendations for Continuous Learning

The IOM report highlighted the need for "continuously learning" and "continuously improving" healthcare organizations. They directly endorse the use of Lean, Six Sigma, and other methods[17] and their more detailed recommendations include many aspects of a Kaizen approach, such as:

- **Recommendation 6:** Care continuity. Improve coordination and communication within and across organizations.
- **Recommendation 7:** Optimized operations. Continuously improve health care operations to reduce waste, streamline care delivery, and focus on activities that improve patient health.
- **Recommendation 10:** Broad leadership. Expand commitment to the goals of a continuously learning health care system.[18]

The IOM elaborates that the characteristics of a continuously learning organization include "a leadership-instilled culture of learning" where a system is "stewarded by leadership committed to a culture of teamwork, collaboration, and adaptability in support of continuous learning as a core aim."[19] The IOM also envisions a learning health system as one in which "complex care operations and processes are constantly refined through ongoing team training and skill building, systems analysis and information development, and creation of the feedback loops for continuous learning and system improvement."[20] That all describes Kaizen and that is becoming the culture of Franciscan and other leading health systems.

The idea of a *learning organization* is not new, as the term has been used for decades by the quality guru W. Edwards Deming and his followers, as well as the *systems thinking* field popularized by MIT professor Peter Senge. Many health systems have found that Lean and Kaizen methods and principles can create a learning organization.[21] In increasingly uncertain times, the organizational ability to learn, improve, and adapt will be helpful, if not absolutely necessary.

Dr. Berwick's Early Call for Kaizen in Healthcare

The word *Kaizen* was introduced to the West by Japanese author and consultant Masaaki Imai in his 1986 book *KAIZEN: The Key to Japan's Competitive Success.* Imai wrote simply, "Kaizen means improvement" and "Kaizen is everybody's business."[22]

Shortly after Imai's book, other healthcare leaders took notice. Dr. Donald M. Berwick is legendary in healthcare quality and patient safety improvement circles, thanks to his advocacy and education work done as the founder and chairman of the Institute for Healthcare Improvement and as the former administrator of the U.S. Centers for Medicare and Medicaid Services. In 1989, Berwick published an article called "Continuous Improvement as an Ideal in Health Care" in the *New England Journal of Medicine*, where he wrote that continuous improvement "holds some badly needed answers for American health care."[23]

Berwick cited Imai with the definition that Kaizen is "the continuous search for opportunities for all processes to get better" and emphasizing that the self-development and the pursuit of completeness found in Kaizen are "familiar themes in medical instruction and history." In highlighting what is different with Kaizen, Berwick criticized disciplinarian-style leaders who look to punish "bad apples" instead of improving processes. He also argued that a leader cannot be "a mere observer of problems," but instead needs to lead others toward solutions.

Berwick highlighted a number of themes, including:

- Leaders must take the lead in continuous quality improvement, replacing blame and finger pointing with shared goals.
- Organizations must invest managerial time, capital, and technical expertise in quality improvement.
- Respect for healthcare professionals must be reestablished, highlighting that they are assumed to be trying hard, acting in good faith; "people cannot be frightened into doing better" in complex healthcare systems.

Berwick's summary of continuous improvement emphasized the culture change required to have everybody work together—removing fear, shame, and finger pointing from the healthcare system. Many organizations post statements about continuous improvement on their websites or on lobby signs. Unfortunately, even in 2012, too many of those statements reflect aspirations rather than reality.

The Impact of Kaizen at Franciscan St. Francis

At Franciscan, the adoption of Kaizen grew gradually, yet impressively, over the first few years since the launch of the Kaizen program in 2007. In the year 2011, 41% of the staff had participated during that year, and 53% of the staff had participated sometime since the launch. In 2011, 82% of all departments had at least one person participate in Kaizen. The growth in employee participation is shown in Figure 1.2, including a slight dip in 2012. Franciscan is working toward having individual participation rates reach 80% in a given year.

As a benchmark, Toyota receives an average of 10 improvement ideas per person each year, after decades of building their Kaizen culture.[24] The number of Kaizens implemented each year at Franciscan is shown in Figure 1.3, at a peak of 1.7 Kaizens per full-time equivalent in 2011. The number at Franciscan is significantly higher than most healthcare organizations and they are working to increase participation each year. In 2013 Franciscan is aiming for 5,000 Kaizens.

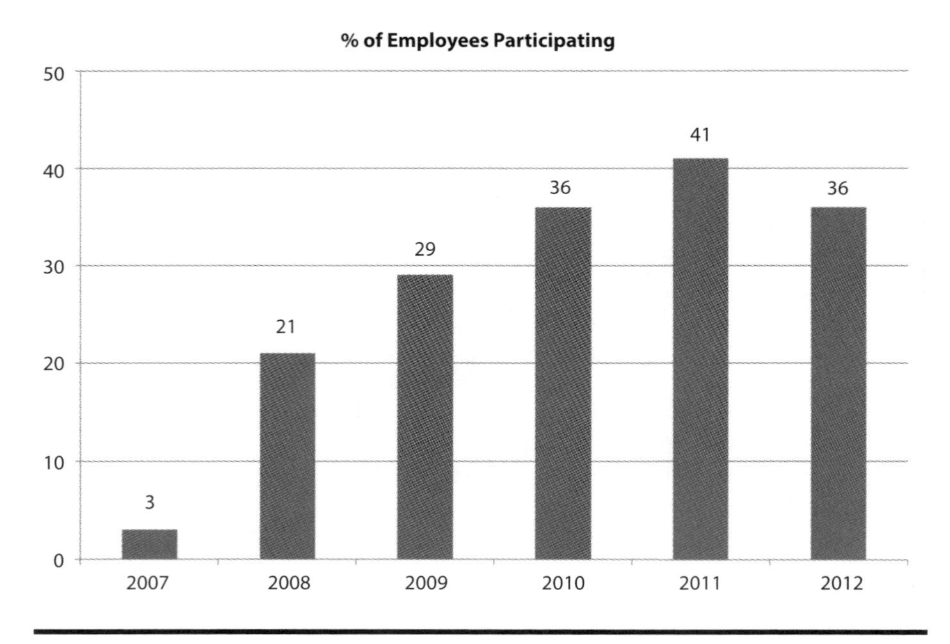

Figure 1.2 Percentage of employees at Franciscan St. Francis Hospitals with a formally submitted Kaizen in each year.

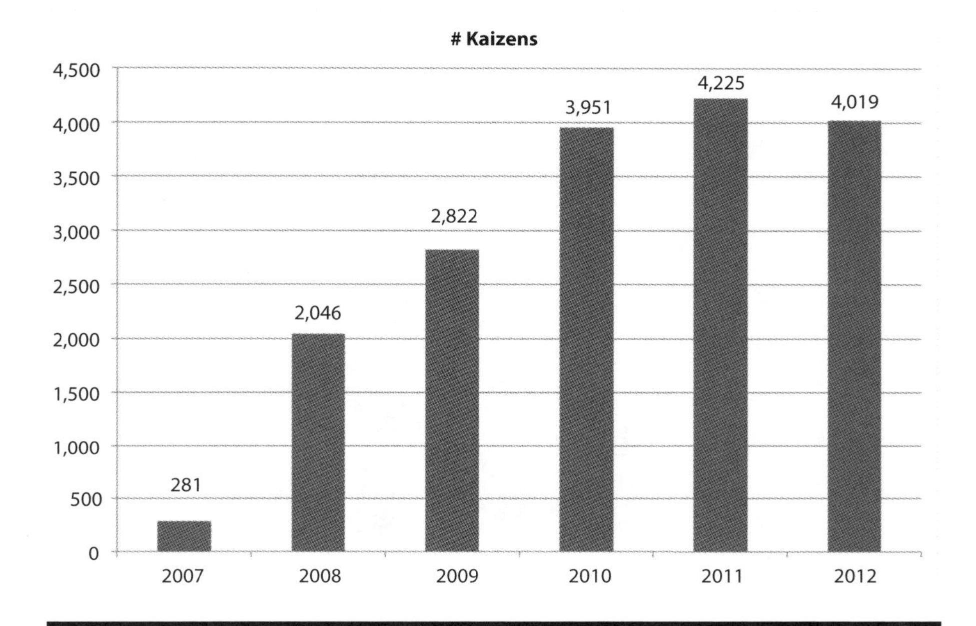

Figure 1.3 Number of formally completed and documented Kaizens each year.

At Franciscan, the Kaizens in 2010 resulted in a total documented savings of over \$3 million. Of that, about \$1.7 million in savings was money that flowed directly to the bottom line, and about \$1.4 million of that was potential dollar savings through, for example, the saving of someone's time. Beyond these documented savings are the benefits from small Kaizens, where it is hard (or not worth the time) to calculate savings. The last 6 years of savings are shown in Figure 1.4. It should be noted that the calculated cost savings were lower in 2011 and 2012, as staff was preoccupied with the installation of a new electronic medical record system and the consolidation of two campuses into one. Basically, leaders did not push for a strong focus on cost savings in those two years, but there is a renewed cost focus in 2013.

Again, bottom-line savings and return on investment (ROI) are not the only things that matter, but they are an important part of the picture for healthcare organizations that are under significant financial pressures. Frontline staff at organizations like Franciscan haven't formally been taught to translate what they do into financial terms. They often don't know the cost of the supplies they use, an hour of their time (fully burdened), or an hour of operating room time. The practice of Kaizen at Franciscan has helped educate staff to get them thinking in financial terms, in addition to (not instead of) everything they already focused on.

The most significant benefit at Franciscan has been the difference Kaizen makes for patients and staff, as this book and the additional examples in our companion book *Healthcare Kaizen* will demonstrate. The softer benefits related to patient

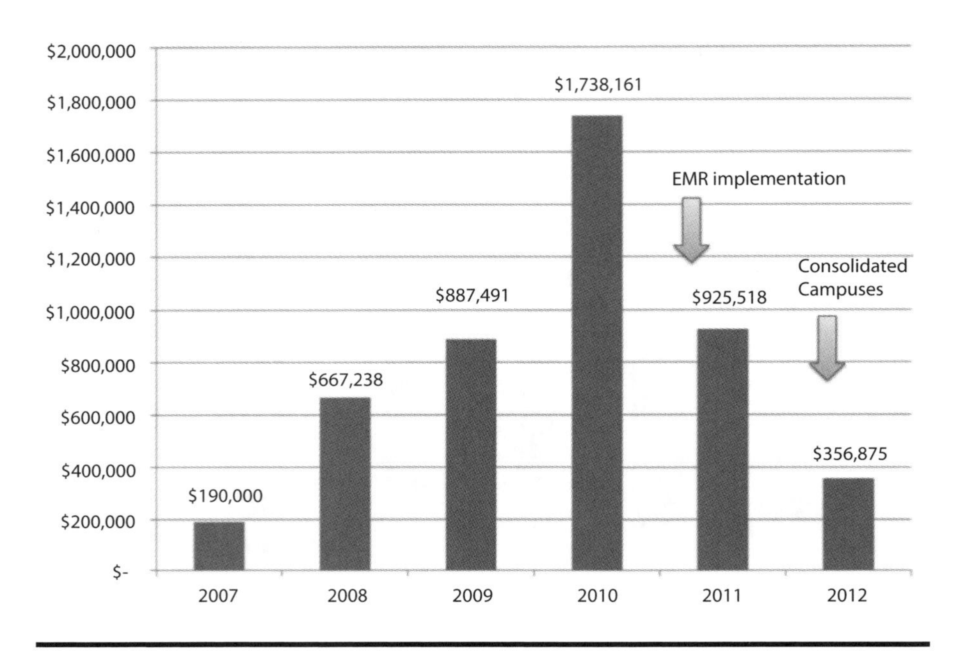

Figure 1.4 Bottom-line "hard" dollar savings for 2007–2012.

safety, outcomes and satisfaction, along with staff safety, learning, and satisfaction, can be hard to quantify but cannot be emphasized enough.

Baptist Health Care (Florida) has a Kaizen program called "Bright Ideas" that was established in 1995 to engage all employees in improvement and innovation. Since 2008, every employee has had the expectation of implementing three ideas per year that will improve patient outcomes, save time, or improve safety. More than 50,000 ideas have been implemented since 2000, and although cost savings is not the primary goal, there has been a total estimated cost savings and avoidance of $50 million. In 2008 alone, almost 14,000 ideas were implemented, or more than two per employee, contributing $10.5 million in cost avoidance and $5.5 million in cost savings.[25] Additionally, full-time voluntary turnover is relatively low, at just 4% annually,[26] and the system has been on the Fortune 100 Best Places to Work for in America twelve of the last fifteen years, including every year from 2003 to 2012.[27]

It Is Not Always about Cost

> The most common types of Everyday Lean Ideas involve shaving time off a process. And when you think about the cost in health care one of our most valuable resources is time.[28]
>
> **—Virginia Mason Medical Center**

The aspect of Kaizen that excites some senior leaders is cost reduction, especially in a tough economic environment. This is very understandable. However, if too much focus is placed on cost savings, staff may get discouraged because they also want to improve quality, safety, and waiting times, while creating a better workplace. One way to address this is for administration to recognize and share Kaizens with varied benefits, not just those that reduce costs. At Franciscan, fewer than 6% of Kaizens are quantified and verified with finance personnel as directly saving bottom-line money for the organization.

> One cannot be successful on visible figures alone … the most important figures that one needs for management are unknown or unknowable, but successful management must nevertheless take account of them.
>
> **—W. Edwards Deming, PhD**

At Goshen, Dague focused improvement efforts on four interwoven areas: customer satisfaction, quality, cost effectiveness, and best people, adding, "If you improve just one area, you're missing the boat. A lot of people I see don't get that." Dague reflected, "You might make a decision that's just driven by finances, like layoffs, but there are all kinds of studies that indicate that if you focus on best people and high-quality patient care, then the money follows. So yes, we have financial goals, but the most important aspect to hitting those goals is happy people."[29]

Several hospitals, including the previously mentioned IU Health Goshen Hospital, ThedaCare, Denver Health, Avera McKennan (Sioux Falls, South Dakota), and Akron Children's Hospital, have a formal "no layoffs" or "no layoffs due to Lean" philosophy,[30] even in the challenging environment in 2012.[31]

These organizations will retrain, reassign, or redeploy any displaced employees to other departments or create opportunities to work in a central Lean or Kaizen group. If employees feared losing their jobs, "Nobody would get very enthusiastic about improvement in that world," says Dr. Dean Gruner, the chief executive and president of ThedaCare.[32]

Another CEO, Gary J. Passema, of NorthBay Healthcare (Fairfield, California), started a Lean program in 2012 as a "better, longer lasting, and less traumatic" way of reducing costs, as previous layoffs were followed by employee counts increasing again and having "savings evaporate" in the following years.

At the Cancer Treatment Centers of America, the objective statement for Kaizen is almost always about improving patient care, says Herb DeBarba, their vice president for Lean Six Sigma. Out of 450 Kaizens they analyzed, only one was financially driven.[33] Bart Sellers, regional manager of management engineering at Intermountain, says "one of the reasons we have been successful [with Kaizen] is that we don't have great expectations about big savings" from improvement ideas.[34]

Another motivation for Kaizen might seem a bit esoteric—learning. Marc Rouppe van der Voort, innovation manager at St. Elisabeth Hospital (Tilburg, The Netherlands), and his colleagues learned, from a Toyota factory visit, that the first priority for Kaizen activity is building problem solving capability and the second priority is the process improvement itself. At St. Elisabeth they mirror the Toyota thinking that "if we learn and there is no improvement, that is good" and "if there is improvement and no learning, that is bad." In 2011 and 2012, St. Elisabeth implemented 4,000 Kaizens through *improvement boards* that are posted in 75 departments.[35]

The Business Case for Kaizen

> We made an investment in developing performance improvement concepts, and it has had lasting impact. The Kaizen movement allows everyone to participate, which is the most exciting and pleasing aspect to our organization, and is responsible for untold small improvements that add up well beyond what we have attributed to the program.
>
> **—Bob Brody**
> *CEO, Franciscan St. Francis Health*

The formal and direct cost of the Kaizen program at Franciscan is very small:

- One day of outside consulting help to get started
- 1.0 full-time equivalent (FTE) in a central Kaizen Promotion Office
- Some designated Kaizen coordinators in different departments, who spend a part of their time on improvement
- A few thousand dollars each year for recognition, incentives, and awards
- The cost of creating and maintaining a database to document and share completed Kaizens

Kaizen does not require major investment in new equipment, facilities, or technologies, making it the perfect strategy for times when money is tight and margins are squeezed. The main cost and investment is the time of staff and leaders at all levels. Kaizen is not complicated, but it requires ongoing effort and discipline. Leaders must often create time for Kaizen by, for example, reducing wasteful meetings. Kaizen frees up more time for Kaizen—if one can get started. As we discuss in Chapter 2, the best way to get started is through small steps.

Rachelle Schultz says that "there is a clear ROI" from their Kaizen and Lean program, as "problems actually get solved now and we track that." Before Kaizen, "the old way was just a constant feed to the black hole, taking up a lot of staff time to talk about problems, but not fix them."[36] She adds, "We break problems down and work toward a long-term goal instead of trying to fix everything overnight. With greater limitations on our staff and physician resources, there is a strong business case for Lean in all we do."

The business case for Kaizen includes small costs and big returns, including:

- Lower staff turnover costs
- Cost reduction and "hard savings"
- Higher revenue and patient throughput
- Cost avoidance and "soft savings"
- Improved quality and patient safety

Lower Staff Turnover Costs

With healthcare facing major shortages of key personnel, including pharmacists, primary care physicians, nurses, and medical technologists, it is imperative to increase employee engagement to help attract and retain people. A 2007 study showed 13% of registered nurses had switched jobs in the first year of their career and 37% "felt ready to change jobs."[37] Research from England shows that 44% of nurses would "leave their job if they could" due to frustrations, including their inability to complete all required nursing tasks during their shift.[38] Kaizen engages staff in fixing problems and reducing the waste that is often at the root of this dissatisfaction, providing a clear path to improving care and reducing turnover.

James Dague preferred to focus Kaizen efforts at Goshen on improving retention, because it can cost $65,000 to replace an employee and much more than that for a nurse or a physician. Increased retention can be the best form of cost reduction, and Dague says, "That's how you can justify honing people's skills to make them more satisfied."[39] Another CEO said, "We don't worry about measuring ROI on Kaizen" because reducing voluntary turnover is justification enough for their program. An AARP study suggests the annual nurse turnover cost for a hospital with 1,000 nurses is $20 million per year.[40]

Cost Reductions and Hard Savings

Many Kaizens do have direct cost savings that can be easily measured and quantified, such as from reducing the use or wastage of supplies or medications. The single Kaizen with the highest value at Franciscan came when three departments worked together to reduce denial of payments from insurance companies. Before, the emergency department (ED) chart did not have enough documentation to avoid denials or write-offs. The team added indicators that showed why a radiology test was run and added another for long-term use of anticoagulants. They surprised themselves with the results, saving the organization over $250,000 a year from their small changes.

CEO Tétreault reports that St. Boniface General Hospital often finds cost reduction opportunities that are uncovered by staff members in the midst of improvement work that is focused on other objectives like quality or waiting times. When working on a project in the cath lab, a staff member asked, "how come we have four types" of a certain supply, which prompted contract consolidation and renegotiation. Tétreault says, "That's a hard and fast financial negotiation, but that's not what we're going after in our transformation." Active participation in Kaizen, "instead of pouring over spreadsheets," means St. Boniface "understands the business better, which allows us to perform better," according to Tétreault.[41]

Sami Bahri, D.D.S., "the world's first Lean dentist," used Kaizen methods to treat the same number of patients in 2006, compared to 2005, but using 40% fewer resources (reducing staff through attrition, not through layoffs). Bahri's office reduced patient waiting times and provided more dental care in a single visit instead of requiring follow-up visits.[42]

Higher Revenue and Patient Throughput

With all of their improvements, Bahri Dental Group added patients and increased their productivity, as measured by *teeth treated per doctor,* by 65% from 2005 to 2012, with the sharpest gains in the last three years. Bahri says, "We did not need to hire more people to treat our patients and the profits went up as well."[43]

Other organizations have used Kaizen as a strategy for increasing patient throughput and revenue. The radiology department at Children's Medical Center Dallas used Kaizen, as part of a larger Lean initiative, to nearly double the number of hours the MRI machines were scanning patients each day, meaning that more patients were treated and their backlog of waiting patients was reduced significantly. Christie Clinic (Champaign, Illinois) built "teamwork and trust" by involving all staff in their improvement efforts, reducing patient wait times by 30% and increasing clinic throughput by 10% over 9 months, while having very high staff satisfaction scores.[44]

Cost Avoidance and Soft Savings

Organizations using Kaizen also find many opportunities for cost avoidance or soft savings. Many Kaizens are focused on saving staff time. Some time savings translate into direct hard savings, because overtime, temporary staff, or overall staff levels are reduced, through attrition or reassignment. Some time savings result in quality or patient satisfaction benefits, which can be difficult to more directly tie to a particular Kaizen. For example, a collection of improvements that free up staff time can lead to reduced patient falls or infections, which has a clear patient benefit and also reduces the amount of unreimbursed care provided by the hospital.

Other soft savings include the cancelation or avoidance of capital purchases or expansion as the result of Kaizen work. One such example is Seattle Children's Hospital (Seattle, Washington), which has avoided $180 million in construction costs by improving throughput and capacity through process improvement.[45]

Improved Quality and Patient Safety

As originally highlighted in Lean Hospitals, Bill Douglas, chief financial officer at Riverside Medical Center, sent a powerful message to their employees early in their Lean journey by saying, "Lean is a quality initiative. It isn't a cost-cutting initiative. But the end result is, if you improve quality your costs will go down. If you focus on patient quality and safety, you just can't go wrong. If you do the right thing with regard to quality, the costs will take care of themselves."[46] Statements like this are very helpful, especially if staff members are cynical about past cost-cutting programs.

Virginia Mason Medical Center (Seattle, Washington) has been widely recognized for using Rapid Process Improvement Workshops (RPIWs), a type of formal weeklong Kaizen project, described more in Chapter 2.[47] Thanks to the improvements that come from its RPIWs, VMMC providers are able to spend more time with patients, leading to better care as well as higher patient and staff satisfaction.[48] Additionally, bedsores were reduced from 8% to 2%, meaning that 838 patients each year avoided a type of harm that is increasingly considered to be preventable.[49] As a result of these quality and safety improvements, the hospital's professional liability insurance rates fell almost 50% from 2004 to 2009, a clear indicator of the program's quantifiable improvements.[50]

Newton Medical Center (Kansas) has recognized the value of employee engagement, saving $1.7 million in a year as the result of 121 ideas that came in after leadership asked people for improvements that would reduce costs or eliminate waste. Val Gleason, the senior VP of physician services, emphasized, "It was not management imposing its will; it was management saying, 'Here's the problem we face, here's the external environment, how are we going to respond to this?'" Furthermore, a hospital spokesperson said, "The ideas allowed them to save money, protect patient care, and protect the integrity of the work force by not having to have any layoffs."[51]

Improvements Have Interwoven Results

Thankfully, the high-level goals, or "true north," of leading healthcare organizations are interwoven. For example, ThedaCare's four true north goals are:

1. Safety/Quality (preventable mortality and medication errors)
2. Customer Satisfaction (access, turnaround time, and quality of time)
3. People (injuries, wellness, and satisfaction)
4. Financial stewardship (operating margin and productivity)[52]

Improvements in these areas can be simultaneous, such as a patient throughput improvement that reduces patient waiting time while improving operating margin. Or, one improvement can lead to another, as indicated by research into healthcare improvement. Research published by the British NHS shows that organizations with higher staff engagement also score higher on measures of financial effectiveness, have higher patient satisfaction, and have lower levels of patient mortality.[53]

As illustrated in Figure 1.5, higher staff engagement correlates highly with lower turnover, better quality, and lower cost. Arguably, one leads to another in a sequential flow.

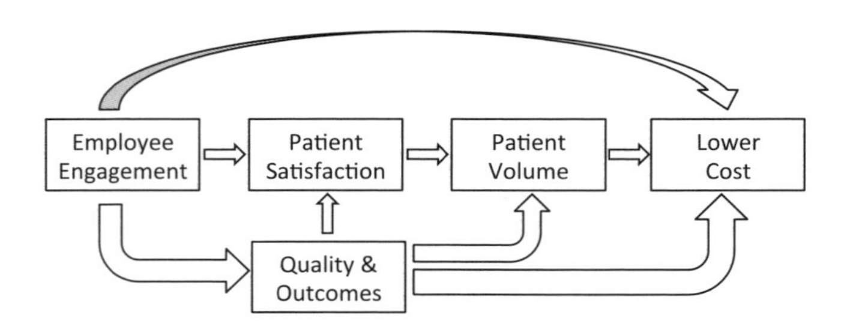

Figure 1.5 Higher staff engagement correlates highly with lower turnover, better quality, and lower cost.

What Executives Need to Do

As Deming said, "Quality starts in the boardroom."[54] Senior leadership is responsible for the management system and organizational culture that either stifles or enables quality and improvement. Culture change and process improvement cannot be outsourced to consultants or delegated to a quality, Lean, or organizational development department.

> This is not the work of the Lean department. It is the work of everyone and it will require a different method of management. It will also require the executives to elevate this to a very high priority level for it to be successful. Lean is not for the faint of heart.
>
> **—David Munch, MD**
> *Chief clinical officer, Healthcare Performance Partners*
> *Former chief clinical and quality officer, Exempla Lutheran Medical Center*

Creating a Management Operating System

To be most successful and sustainable, Kaizen should be more than a standalone program. Toussaint says senior leaders must create a coherent and holistic "operating system" for the organization,[55] with Kaizen being a "critical part," along with other Lean management practices.[56] CEOs must "get personally involved and change their behaviors," as opposed to just hiring a consultant to facilitate some improvement events, says Toussaint. Chapter 7 details additional senior leadership behaviors that need to be practiced and modeled for others in the organization.

Following on Toussaint's work, ThedaCare created a "business management system" that spelled out consistent management activities and behaviors for leaders at all levels, in order to "achieve and sustain continuous daily improvement."[57] Many healthcare organizations are embracing the holistic "Shingo model" for operational excellence that connects guiding values and principles with continuous improvement concepts into a system that aligns the entire organization to achieve the highest possible results.[58] Changing culture requires changing the way leaders at all levels behave each and every day.

> We have a wide variation in how we are managing healthcare organizations and that is leading to a lot of [quality] defects.[59]
>
> **—John Toussaint, MD**

Tying Kaizen to Strategy

The focus of Kaizen, especially in the early stages of an organization's adoption, is often on issues that "bubble up" from frontline staff and the things they notice or struggle with in the course of their daily work. This staff-driven improvement might raise concerns of having a thousand random Kaizens that may conflict with each other, suboptimize one department, or have little impact on the organization's goals and targets.

John Shook, CEO of the Lean Enterprise Institute, teaches that this approach is neither completely top down nor completely bottom up. Strategies and goals flow, generally, in a top-down direction, while ideas generally flow upward from front-line staff.[60] In both directions, there is a "catch ball" process of collaboration, where strategies are adjusted based on input from other levels and improvement ideas are refined based on input from leaders.[61]

ThedaCare and St. Boniface are among those who use the Lean model of *strategy deployment* for planning and measuring performance and improvement.[62] The focus and prioritization of Kaizen is guided by the organization's strategy, including its true north goals and annual objectives. Shana Herzfeldt, a medical services unit manager at ThedaCare, says, "As a manager, I'm able to see on a daily basis my business, understand what's impacting my metrics, and not be surprised at the end of the month. As for my staff, they feel engaged and empowered knowing that their decisions greatly impact the work they do on a daily basis."[63]

Connecting Kaizen to the Mission

Beyond the business case, more healthcare organizations are embracing Kaizen methods because the methodology aligns with their strong sense of purpose and mission. Franciscan's Kaizeneers are taught to remember, "we, not me." This can help connect what you are doing to the organization's mission to serve patients and the community.

Kaizeneer: Franciscan uses the term *Kaizeneer* for staff members who practice Kaizen—a term that is a combination of the words *Kaizen* and *engineer.* Engineers are designers, and those who do Kaizen are essentially designing or redesigning their world around them. If Disney has Imagineers who design their theme parks, why can't healthcare have Kaizeneers who redesign the healthcare work environment?

In Mark's experience working with Children's Medical Center Dallas, leaders in the laboratory and radiology very naturally talked about how improvement efforts needed to tie back to their mission, informally stated as "taking care of kids." Clay York, laboratory operations manager, frequently emphasizes tying improvements back to the patients. Even though the lab is physically disconnected from patients, York strengthens the emotional connection to their unseen patients by talking about how turnaround times affect the ability to get children discharged and back to school. York asks, "Are we doing this for the performance measures or for the kids?" Clearly, they are doing it for the kids, which leads to improved measures.

At IU Health Goshen Hospital, every meeting starts with the reading of the organization's mission or a department's mission, says Dague, along with some reflection on how well they are living those values. Dague and the other leaders help individuals connect their improvement work to their personal mission, saying:

> One of the things we talk about in healthcare is one's personal mission. Why did you get into this job and into this profession? How are you doing in completing your life's mission? We want to get out of your way, in our culture, to allow you to fulfill that mission. That's something that brings this home to the individual far more than I ever thought it would.[64]

Conclusion

Masaaki Imai more recently defined Kaizen as "everyday improvement, everybody improvement, and everywhere improvement."[65] At Franciscan, *everybody* means clinical and nonclinical staff and leaders at all levels—including the CEO, COO, and other top leaders who directly participate in Kaizen activities.

Both coauthors have seen firsthand and believe strongly that healthcare professionals at all levels:

- care deeply about their patients,
- want to provide the highest-quality ideal care to each patient, and
- have the ability and the desire to use their creativity to improve their workplace.

Healthcare has a 100-year-long track record of trying to adopt and emulate quality improvement methods, including Total Quality Management, Continuous Quality Improvement, Six Sigma, and Lean. Yet, daily continuous improvement seems to be more a goal than a reality in a majority of healthcare organizations.

Why don't we have more improvement? Rather than pointing fingers at individuals—frontline staff, managers, or senior leaders—everyone should work to understand the systemic barriers and the oft-unspoken mindsets that interfere with making continuous improvement a reality. Your organization may have tried other improvement methodologies in the past, including Total Quality Management and Six Sigma. If past attempts at continuous improvement did not work out, it might be helpful to stop and reflect upon the systematic root causes of those struggles before moving forward with this book or with Kaizen.

Kaizen should not be just a one-time flurry of ideas, nor should it be just a one-time reaction to an organization facing financial pressures, as are rampant today. A so-called "burning platform" or crisis might prove motivating to some, but the pressure of a crisis might also harm creativity and have people hold back ideas if they fear they could be associated with job cuts that might occur in a tough economic environment. Ideally, the crisis would be an opportunity to learn and practice Kaizen methods that would continue even after the crisis has subsided. Kaizen, as a part of Lean, should be part of an organizational strategy and comprehensive management system.

Discussion Questions

- If your organization has tried other improvement programs that did not work or sustain, what are some of the root causes of that failure?
- What are some reasons that Kaizen has not been embraced more widely in healthcare over the past 20 years? What are some of the specific reasons within your organization?
- What is a single small Kaizen that you can identify and implement today in your own work?
- What do you want to accomplish through Kaizen? How do you strike the proper balance in talking about benefits to patients and staff, which are sometimes hard to quantify, and cost savings or other financial benefits?
- What is your "case for change" or "burning platform" for improvement? Does everybody in your organization understand this? If not, how can we educate them and inspire them to improve?

Endnotes

1 Brody, Bob, personal interview, January 2013.
2 Kato, Iaso, and Art Smalley, *Toyota Kaizen Methods: Six Steps to Improvement* (New York: Productivity Press, 2010), 102.
3 Jacobson, Gregory H., N.S. McCoin, R. Lescallette, S. Russ, and C.M. Slovis, "Kaizen: A Method of Process Improvement in the Emergency Department," *Acad. Emerg. Med.* 16, no. 12 (2009): 1341–1349.
4 York, Clay, Personal interview, July 2011.
5 IU Health Goshen Hospital, "GHS Colleagues Meet Challenge; CEO Shaves Head," January 1, 2010, http://iuhealth.org/newsroom/detail/ghs-colleagues-meet-challenge-ceo-shaves-head/ (accessed January 12, 2013).
6 Dague, James, personal interview, July 2011.
7 Goshen News, "Health System among Best Places to Work in Indiana," March 12, 2008, website, http://goshennews.com/business/x395807567/Goshen-Health-a-Best-Place-to-Work-for-third-year (accessed January 12, 2013).

8 Institute of Medicine, "Report Brief: Best Care at Lower Cost—The Path to Continuously Learning Health Care in America," September 2012, 2, http://www.iom.edu/~/media/Files/Report%20Files/2012/Best-Care/BestCareReportBrief.pdf (accessed January 5, 2013).
9 Institute of Medicine, Full Report, 2012, 105.
10 Campbell, Dennis, "Diabetes Threatens to 'Bankrupt' NHS within a Generation," *The Guardian*, April 25, 2012, http://www.guardian.co.uk/society/2012/apr/25/diabetes-treatment-bankrupt-nhs-generation (accessed January 5, 2013).
11 Smith, Rebecca, "Patient Care Will Suffer under Budget Cuts Warn NHS Finance Bosses," *The Telegraph*, September 27, 2012, http://www.telegraph.co.uk/health/healthnews/9568068/Patient-care-will-suffer-under-budget-cuts-warn-NHS-finance-bosses.html (accessed January 5, 2013).
12 Smith, "Patient Care Will Suffer."
13 Rahmani,Tabassum, "Under Budget Pressure, Canada Slowly Rethinks Health Care Model," *Heartland*, n.d., http://news.heartland.org/newspaper-article/under-budget-pressure-canada-slowly-rethinks-health-care-model (accessed March 3, 2013).
14 Tétreault, Michel, personal interview, January 2013.
15 Institute of Medicine, 4.
16 Robbins, Anthony, *Awaken the Giant Within: How to Take Immediate Control of Your Mental, Emotional, Physical, and Financial Destiny* (New York: Simon & Schuster, 1992), 97.
17 Institute of Medicine, Full Report, 4–7.
18 Institute of Medicine, Full Report, S-20.
19 Institute of Medicine, Full Report, 5.
20 Institute of Medicine, Full Report, 5.
21 Adams, Jim, and Mark Graban, "CMCD's Lab Draws on Academics, Automakers, and Therapists to Realize Its Own Vision of Excellence," *Global Business and Organizational Excellence*, May/June 2011, 17, doi: 10.1002/joe.20383.
22 Imai, xxix.
23 Berwick, D.M., "Continuous Improvement as an Ideal in Health Care," *New England Journal of Medicine* 320, no. 21 (1989): 1424–1425. All references to Berwick in this chapter are to this article.
24 Liker, Jeffrey K., and David Meier, *The Toyota Way Fieldbook* (New York: McGraw-Hill, 2006), 261.
25 Brophy, Andy, and John Bicheno, *Innovative Lean* (Buckingham, England: PICSIE Books, 2010), 138.
26 CNN Money, "100 Best Companies to Work For," 2012, website, http://money.cnn.com/magazines/fortune/best-companies/2012/snapshots/42.html (accessed January 12, 2013).
27 Baptist Health South Florida, "Awards and Recognition," http://baptisthealth.net/en/about-baptist-health/Pages/Awards-and-Recognition.aspx (accessed January 12, 2013).
28 Virginia Mason Medical Center, "Using Lean Ideas in Our Everyday Work," http://virginiamasonblog.org/2013/01/30/using-lean-ideas-in-our-everyday-work/ (accessed January 30, 2013).
29 Dague, interview.
30 Graban, Mark, "How Lean Management Helped Hospitals Avoid Layoffs," FierceHealthcare.com, October 1, 2010, http://www.fiercehealthcare.com/story/how-lean-management-helped-hospitals-avoid-layoffs/2010-10-01 (accessed January 12, 2013).

31 Graban, Mark, "Lean as an Alternative to Mass Layoffs in Healthcare," *Becker's Hospital Review* website, September 6, 2012, http://www.beckershospitalreview.com/hospital-management-administration/lean-as-an-alternative-to-mass-layoffs-in-healthcare.html (accessed January 12, 2013).
32 Boulton, Guy, "ThedaCare's 'No Layoffs' Practice May Improve Firm," *Journal Sentinel Online*, December 20, 2008, http://www.jsonline.com/business/36477214.html (accessed January 12, 2013).
33 DeBarba, Herb, personal interview, July 2011.
34 Sellers, Bart, email correspondence, August 29, 2011.
35 Graban, Mark, "Lean Healthcare Transformation Summit 2012, Day 1," http://www.leanblog.org/2012/06/lean-healthcare-transformation-summit-2012-day-1/ (accessed January 30, 2013).
36 Graban, Mark, "Podcast #164–Rachelle Schultz, CEO of Winona Health." LeanBlog.org, http://leanblog.org.164 (accessed May 5, 2013).
37 American Association of Colleges of Nursing, "Nursing Shortage Fact Sheet," website, http://www.aacn.nche.edu/media/factsheets/nursingshortage.htm (accessed January 12, 2013).
38 *The Evening Standard*, "Nurses Struggle to 'Get the Job Done' Due to Staff Shortages and Lack of Time on 12-Hour Shifts," website, http://www.standard.co.uk/news/health/nurses-struggle-to-get-the-job-done-due-to-staff-shortages-and-lack-of-time-on-12-hour-shifts-7975862.html (accessed January 12, 2013).
39 Dague, James, personal interview, July 2011.
40 AARP, "What Are the Costs of Employee Turnover?" website, http://www.aarp.org/work/employee-benefits/info-04-2011/what-are-the-costs-associated-with-employee-turnover.html (accessed January 5, 2013).
41 Tétreault, Michel, personal interview, January 2013.
42 Bahri, Sami, *Follow the Learner* (Cambridge, MA: Lean Enterprise Institute, 2009), 85.
43 Bahri, Sami, personal interview, February 2013.
44 Toussaint, John, "A Management, Leadership, and Board Road Map to Transforming Care for Patients," *Frontiers of Health Service Management*, March 2013, http://www.createvalue.org/data/blog/Frontiers_29_3%20Spring-FINAL_Toussaint%20exec%20summary.pdf (accessed May 5, 2013), 8.
45 Weed, Julie, "Factory Efficiency Comes to the Hospital," *NYTimes.com*, July 11, 2010, http://www.nytimes.com/2010/07/11/business/11seattle.html?_r=2&emc=eta1&pagewanted=all (accessed January 12, 2013).
46 Graban, Mark, "Riverside Medical Center Puts Lean in the Laboratory," *Society of Manufacturing Engineers Lean Manufacturing Yearbook 2007* (Dearborn, MI: Society of Manufacturing Engineers), 56.
47 Shigujitsu USA, "Interview with Dr. Gary Kaplan, Virginia Mason Medical Center," website, http://www.shingijutsuusa.com/testimonials.html (accessed January 12, 2013).
48 Virginia Mason Medical Center, "VMPS Facts," website, 2010, https://www.virginiamason.org/workfiles/pdfdocs/press/vmps_fastfacts.pdf (accessed January 12, 2013).
49 Brophy, Andy, and John Bicheno, *Innovative Lean* (Buckingham, England: PICSIE Books, 2010), 141.
50 Virginia Mason Medical Center, "Innovation in Health Care," website, http://www.createvalue.org/docs/case-study-virginia-mason.pdf (accessed January 12, 2013).

51 Shideler, Karen, "Staff's Ideas Save Money at Newton Hospital," *Wichita Eagle*, November 24, 2009, http://www.kansas.com/2009/11/24/1068956/staffs-ideas-save-money-at-newton.html#ixzz1N1kSo77u (accessed January 12, 2013).
52 Barnas, Kim, "ThedaCare's Business Performance System: Sustaining Continuous Daily Improvement through Hospital Management in a Lean Environment," *Joint Commission Journal on Quality and Patient Safety* 37, no. 9 (2011): 390.
53 NHS Employers, "The Staff Engagement Challenge: A Factsheet for Chief Executives," http://www.nhsemployers.org/Aboutus/Publications/Documents/23705%20Chief-executive%20Factsheet_WEB.pdf (accessed January 25, 2013).
54 Deming, W. Edwards, Joyce Orsini, and Diana Deming Cahill, *The Essential Deming: Leadership Principles from the Father of Quality* (New York: McGraw-Hill, 2012), 39.
55 Toussaint, *Frontiers*, 3.
56 Toussaint, John, personal interview, January 2013.
57 Barnas, "ThedaCare's Business Performance System," 387.
58 ThedaCare Center for Healthcare Value, "Assessing and Accelerating Your Lean Transformation," http://www.createvalue.org/docs/Assessment_HVN_December2011 (R5).pdf (accessed January 27, 2013).
59 Toussaint, John, personal interview, January 2013.
60 Shook, John, presentation, First Global Lean Healthcare Summit, June 25, 2007.
61 Graban, Mark, *Lean Hospitals: Improving Quality, Patient Safety, and Employee Engagement*, 2nd edition (New York: Productivity Press, 2012), 180.
62 Graban, *Lean Hospitals*, 181.
63 ThedaCare Center for Healthcare Value, *Thinking Lean at ThedaCare: Strategy Deployment*, DVD, 2011.
64 Dague, James O., personal interview, July 2011.
65 Graban, Mark, "Masaaki Imai Kaizen Video: Share Kaizen Stories for My Upcoming Book," http://www.leanblog.org/2011/06/masaaki-imai-kaizen-video-share-kaizen-stories-for-my-upcoming-book/ (accessed January 30, 2013).

Chapter 2

What Is Kaizen?

Quick Take

- Kaizen does not mean change, it means "change for the better," or improvement. A change is an improvement only when things are made better in some way.
- Kaizen starts with small changes, often the smallest possible.
- Kaizen follows Plan-Do-Study-Act (PDSA) and the scientific method.
- To build a culture of continuous improvement, it is important to not punish people for "failed" improvement attempts.
- Kaizen is one of the two pillars of the Lean management system.
- Kaizen is very different than the old, mostly failed, suggestion box model because it is faster, more transparent, and more collaborative.
- Our people are the only real differentiator or competitive advantage for an organization.

> Kaizen is about changing the way things are. If you assume that things are all right the way they are, you can't do Kaizen. So change something![1]
>
> —**Taiichi Ohno**
> *Co-creator of the Toyota Production System*

Bubbles for Babies

Hope Woodard, an ultrasound tech at Franciscan St. Francis Health, noticed that her young patients were often uncomfortable when she pressed the cold hard ultrasound probe to their skin. The children had difficulty staying still during the

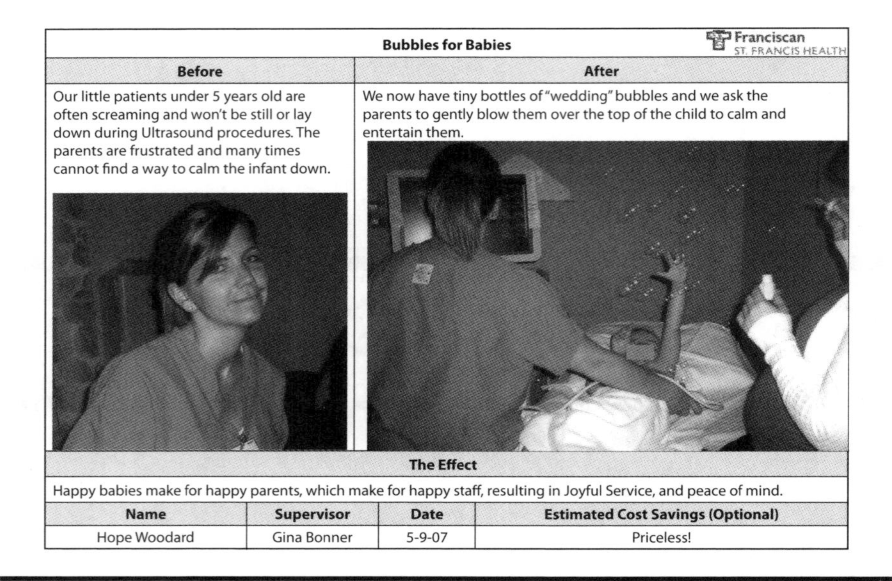

Bubbles for Babies

Franciscan ST. FRANCIS HEALTH

Before	After
Our little patients under 5 years old are often screaming and won't be still or lay down during Ultrasound procedures. The parents are frustrated and many times cannot find a way to calm the infant down.	We now have tiny bottles of "wedding" bubbles and we ask the parents to gently blow them over the top of the child to calm and entertain them.

The Effect

Happy babies make for happy parents, which make for happy staff, resulting in Joyful Service, and peace of mind.

Name	Supervisor	Date	Estimated Cost Savings (Optional)
Hope Woodard	Gina Bonner	5-9-07	Priceless!

Figure 2.1 A simple Kaizen that improved staff and patient satisfaction.

procedures, so parents would get frustrated and often could not find a way to calm their child down.

Hope thought about how she could create a better experience for her little customers. She brought in a small bottle of bubbles from a wedding she had attended and asked the parents to gently blow them over the top of the child to calm and entertain them, as documented in Figure 2.1. As she expected, the bubbles kept the babies calm for their procedures. Happy, distracted children allowed Hope and the other techs to more quickly capture better-quality images for the radiologists, making it better for the patient, for the parents, and for the ultrasound techs. In the course of making her job easier, Hope also added value to the customer experience.

The best Kaizens have several things in common. They increase customer and patient satisfaction while, at the same time, improve the productivity and the quality of healthcare delivery. Small, low-cost improvements can indeed make a difference to patients while increasing the pride and joy felt by healthcare professionals. It created more joy in her work life every time Hope used bubbles for babies.

Kaizen = Continuous Improvement

The word Kaizen generally means "change for the better." These changes can include team projects, such as *Kaizen Events* or *Rapid Improvement Events* (RIEs), as discussed later in Chapter 3. But, as Masaaki Imai emphasizes, Kaizen should be practiced by everybody, everywhere, every day.[2]

In this book, we will use the term Kaizen in the context that is less common, and perhaps less appreciated in healthcare (as well as other industries)—continuous improvements that happen without the formal structure of a large team or a major project. ThedaCare refers to this Kaizen and PDSA process as Continuous Daily Improvement (CDI). ThedaCare conducts weeklong RIEs, but they also use daily Kaizen methods to reach their goal of every person being a problem solver each and every day.[3]

Imai's core Kaizen concepts were summarized in a 2009 article written by Gregory Jacobson, MD, and others for the journal *Academic Emergency Medicine* that described key Kaizen mindsets in a hospital setting, including the following:[4]

- Continually improve, with no idea being too small.
- A major source of quality defects is problems in the process.
- Focus change on commonsense, low-cost, and low-risk improvements, not major innovations.
- All ideas are addressed and responded to in some way.
- Collect, verify, and analyze data to enact change.
- Empower the worker to enact change.

Kaizen Is Not Just Change, It Is Improvement

With Kaizen, we want more than a lot of activity and change; we really want improvement and learning. Improvement comes when we can state that things have been made better in one or more dimensions, including safety, quality, productivity, or having a less-frustrating workplace. Not all changes are improvements. For example, a change to a process that makes it harder for nurses to gather the supplies needed to start an IV would likely not be considered an improvement, because it would delay patient care and cause more work for the nurses.

Kaizen involves finding a better way to do the work, not cutting corners. If a staff member wants to eliminate a step from their work, they should talk to their coworkers to ensure that the change does not negatively impact the patient or someone else in the overall process.

A possible improvement should be proposed as a hypothesis to be tested in practice. For example, a materials management team might propose, "If we rearrange the clean utility room to stock items in the order of their computerized order number, then it reduces the amount of time required to restock rooms each day." After testing that change for three days, the materials management team might conclude that they, indeed, saved them 30 minutes per day. As their test confirmed their expectations, the materials team might decide to share this change with other units.

Large, complex organizations, such as those in healthcare, need to be aware that one area's improvement may cause side effects in other areas. For example, if the change helped materials management but made work more difficult for the nurses who take items from the supply rooms, then, when looking at the big picture, the change might not be an improvement after all.

Kaizen Starts with Small Changes

At first, Kaizens tend to be small, local changes. In many organizations, the focus of improvement is on innovation or larger scale improvements. An organization might traditionally focus energy on one major initiative or innovation, such as a new building or a new electronic medical record (EMR) system, which gets positioned as the solution to most of the hospital's problems, but often falls short of delivering on all its promises. Furthermore, new service or product offerings often don't become a hit on their first debut and need to be rethought or improved upon over time.

A Kaizen organization supplements large, strategic innovations with lots of small improvement ideas—the equivalent of singles and doubles, instead of only hitting home runs, to use a baseball analogy. The expectation is that a large number of small changes lead to an impressive improvement in an organization's core measures. Small changes, which can be completed more quickly than major projects, can build enthusiasm and problem-solving skills that people can then apply to larger problems.

> There are no big problems, there are just a lot of little problems.
>
> **—Henry Ford**

In many large organizations, employees can feel intimidated by the overwhelming number of people with whom they must coordinate to make large-scale improvements. Kaizen encourages employees to start with small changes that do not require coordinating with a large number of people. In a Kaizen approach, we do not start by trying to improve what others do. Instead, we start by improving what we do individually. Once benefits accrue from a few small improvements, motivation and confidence will grow, allowing people to tackle more difficult, more time-consuming improvements. The best way to get started is to make it quick and easy and then just do it.

UCLA psychologist Robert Maurer, PhD, has written about how large changes cause an instinctive "fight or flight" response in humans, causing people to become less creative as the higher-thinking part of the brain shuts down out of fear. Maurer's patients are asked to start exercising by walking in place for just one minute instead of being asked to suddenly start working out 30 minutes every day.[5] Exercising for 60 seconds is less scary, so patients actually take action, gradually working their way up to longer, more meaningful workouts, as they build enthusiasm and confidence from those initial small steps. Starting small is better than not starting at all.

A Small Kaizen with Great Meaning

At Riverside Medical Center (Kankakee, Illinois), support staff would sometimes face the awkward situation of entering a patient room to find a grieving family with a patient who had just passed away. During an initial Kaizen program, Darlene, a member of the housekeeping team, made a simple yet effective and beautiful suggestion to prevent that from occurring again. She created an angel sign that could be placed on the door when a patient passed away. Ancillary departments were instructed to look for the sign so they could remain respectful of the deceased and their family. The sign was also a subtle way to maintain privacy and dignity for the families, because other visitors might just think the sign was a decoration.

Kaizen Involves the People Who Do the Work

A common workplace cliché, often used by managers, is that "people hate change." The expression often goes unchallenged, as leaders use it, perhaps to complain that people will not do what the boss wants them to do. With Kaizen, we learn that people love their own ideas and they love change when they initiate it. The great management thinker Peter Scholtes said simply, "People do not resist change, they resist being changed."[6]

Kaizen reduces some of this "resistance" because leaders get everybody involved in continuous improvement—improving work that matters to them and to their patients. Kaizen is not an approach that is limited to managers or improvement specialists from a central department. Kaizen is for everyone. Once introduced to the approach, a great majority of people are excited to initiate opportunities to apply Kaizen in their work lives and even in their personal lives at home, such as improving their morning routine in getting ready for work.[7]

James Dague, retired CEO of IU Goshen Hospital, emphasizes that you have to involve the people who actually do the work every day in improvement activities. Dague recalls how he previously worked in health systems where engineers would come in for a day to observe in a nursing unit, "generalizing the staffing patterns" based on that one day, and he has "never seen that work well." He adds, "I think not involving people, having changes are going to be imposed on you, will get you a lot of resistance and negativity right off the bat."

Kaizen, PDSA, and the Scientific Method for Improvement

Plan-Do-Study-Act (PDSA), sometimes referred to as Plan-Do-Check-Act (PDCA), is an iterative learning, improvement, and problem-solving model based on the scientific method.

The PDSA steps are:

Plan: Initiating a change by understanding the current situation and root cause of problems; developing a change and stating a hypothesis about what will occur with the change

Do: Carrying out a small-scale test, or pilot, of the change

Study: Testing the change and its hypothesis: gathering data, observing the changes and outcomes

Act (or Adjust): Based on those results, deciding to accept, adopt, and spread the change, or making adjustments (or trying something different)

PDSA is an iterative model, so a successful change leads us to a new starting point for continued improvement. Our process for spreading a change will include testing it in a larger area or in different conditions, learning from each successive cycle. Leaders should not expect people to solve a large and complex problem with a single Kaizen. Each successive small improvement should be celebrated, rather than grousing about why the whole problem was not solved at once. Paraphrasing Voltaire, we cannot let perfect be the enemy of better.

> Our own attitude is that we are charged with discovering the best way of doing everything, and that we must regard every process employed in manufacturing as purely experimental.[8]
>
> **—Henry Ford**

We Often Succeed as the Result of Failing More

In many organizations, a change that does not meet improvement goals or targets (or an idea that just flat out does not work as expected) might be considered a *failure.* If the Study phase shows that a change was not really an improvement, we need an environment where people are not punished for their attempts at Kaizen. If we have fear and punishment, people will be afraid to suggest changes or they will become incredibly cautious, only proposing those things that are certain to work. As the University of Michigan's Mike Rother says, "The idea is to not stigmatize failures, but to learn from them."[9]

Even without the fear of punishment, some people are afraid that they will be embarrassed by trying something unsuccessfully. Instead of PDSA, we might observe the following dysfunctional cycles in an organization:

P-D: Plan-Do—not studying to see if the change was really an improvement; just assuming things are better as the result of a change

P-D-J-R: Plan-Do-Justify-Rationalize—knowing, but being unwilling to admit, that our change did not lead to improvement

It is sometimes said that Toyota is successful on a larger scale because they have a high tolerance for failure in small improvement initiatives. In the long run, using a failure as an opportunity to learn creates a stronger organization.

> Failure is only the opportunity to begin again more intelligently.
>
> **—Henry Ford**

"Failure" Should Result in Learning

Some small changes have a clear, indisputable benefit. For example, a laboratory medical technologist rearranges supplies and equipment on her workbench so that the most frequently used items are at arm's reach instead of being buried in ankle-height drawers. This change saves time and improves ergonomics, leading to faster test results for the patient—this is a change for the better.

Later on, this same technologist might decide, unilaterally, to run a certain low-volume test just one day a week instead of once each day. The technologist's idea is to save the waste of unused reagents in a test pack, as the kit costs the same whether it is used for one test or three. The technologist is trying to save money by making better use of each kit. Unfortunately, this local cost savings might not be a change for the better of the overall system if the batching of the tests causes delays in medical decision making or extends a patient's length of stay.

In a culture that embraces Kaizen principles, this sort of "failure" is seen as a learning opportunity for individuals and the organization, so people can understand the broader impact of their improvements. Leaders need to recognize the effort and desire to improve while teaching people how to make better improvements in the future. The Kaizen approach to management requires that leaders' daily actions encourage ongoing improvement rather than stifle it.

> Fall seven times. Stand up eight.
>
> **—Old Japanese Proverb**

Changing Back Can Be Better for Babies

In early 2008, the Franciscan maintenance department replaced the manual paper towel dispensers in the NICU with hands-free automatic paper towel dispensers. One automated dispenser located near a group of babies made a loud grinding noise each time it dispensed a paper towel and nurses noticed babies flinching when this happened. Occasionally, the noise would wake one of the babies and the nurses knew how important rest was for recovery. Most of Paula Stanfill's nurses chose a career in the NICU because of their passion and compassion for babies.

After some debate, her nurses suggested they go back to the manual dispensers. Paula wondered if she should let them because it seemed as though they were going backward. Then, her staff measured the decibel level of the automatic paper towel dispenser and found it was greater than 50 decibels. Paula was convinced. She approved having the automatic dispenser replaced with the old manual one. It was not as fancy, but it was better for the babies under their care. The babies were happier and healthier, which led to happier staff, which made Paula happy.

Prior to this change, Kaizen was being slowly embraced by her department. A few weeks later, she noticed the Kaizen program growing more rapidly. When she questioned her staff, they told Paula that she had demonstrated that it is okay to try things and fail and that they can go back if they needed to. Paula had sent the message that it is safe to test ideas. Paula learned that she needed to listen carefully to her staff and that sometimes going back is the way to go forward.

> We know we've turned the corner [with Kaizen] when staff get excited about a PDSA test failing.
>
> **—Ray Seidelman**
> *Manager of Performance Improvement, Iowa Health System*

Kaizen Is Not a Suggestion System—It Is an Improvement System

Sometimes when people hear about Kaizen, they say, "We already do that—we have a suggestion box!" Suggestion box systems date back more than 100 years. In 1892, National Cash Register Company (NCR) was the first U.S.-based company to implement a companywide employee suggestion program described as the "hundred-headed brain" by their CEO, John Patterson. While Patterson realized that employees had valuable ideas, the system fell into disuse over time as new leaders took over. The system did not fail because workers "stopped having good ideas," but rather because "the predominant view of the workplace became one in which managers were expected to do the thinking and workers were expected to do what they were told."[10]

Many industries, including manufacturing, are still recovering from the outdated notion that working and thinking should be separated, an idea popularized by Frederick W. Taylor (1856–1915), one of the fathers of industrial engineering. As an early efficiency expert, Taylor stood over workers, timing them and devising ways for them to do their work differently. Although many of Taylor's analysis and work improvement methods, such as time and motion studies, are still in use today, the difference is in the mindset and how these methods are applied. Today, people are analyzing and improving their own work.[11]

Ray Seidelman emphasizes that Kaizen is not a suggestion system, where proposals might include "I need another computer." Rather, it is an approach where people identify problems that get in the way of ideal care. An example might be a staff member saying, "I couldn't find the information I needed about this medication in the record," and then responding by following the scientific method for improvement.[12]

Common Dysfunctions of Suggestion Systems

While well intended, the classic suggestion box has more downsides and dysfunctions than success stories. The problem is not the box itself, but the way it is managed (or ignored). Kaizen and modern improvement systems are based on a different philosophy. Instead of putting the entire approval burden on supervisors, Kaizen is much more collaborative. Kaizen principles could be applied to suggestion boxes, but there are different mechanisms that are faster, more transparent, and more collaborative, than boxes. These methods include bulletin boards and software systems, as described in Chapter 5.

> Many companies assume that the failure of the suggestion box approach is with employees that don't care, but if we dig a little deeper we find it is the system itself that squashed enthusiasm.[13]
>
> **—Bruce Hamilton**
> *President of GBMP*

Suggestion Boxes Are Rarely or Never Opened

At one hospital Mark worked with in the early days of a Lean journey, he found a suggestion box on the wall of the laboratory. One of the team members left to go find the key to the locked box, returning 20 minutes later to announce that nobody could find the key! This story became a very clear illustration to the laboratory staff and leaders of why the suggestion box did not work and how their new approach to Kaizen would have to be different.

> Why is there a lock on your suggestion box? Are you afraid that the competing hospital across the street is sending people over to steal your good ideas?
>
> **—Mark Graban**

Suggestion Box Systems Are Slow, with Poor Feedback

Suggestions often sit for weeks or months in a locked box. Employees are often not happy with the slow or inadequate communication they get from managers once the box is opened. Kaizen emphasizes implementing most ideas, or at least giving timely and collaborative feedback to every employee who has an idea, not just those that are deemed accepted by a far-off committee.

Too Many Suggestions Are Rejected or Ignored

In suggestion systems, people get discouraged when their ideas are rejected, leading them to, understandably, stop participating. In a Kaizen approach, an idea is the starting point for a dialogue between employee and supervisor, working together to understand the real problem to be solved. Often, the original idea, even if not deemed practical, sparks a new idea that *can* be implemented. When it is said that Toyota implements more than 90% of their employee ideas, that means that they found *something* to implement, not necessarily the original ideas as presented. Some healthcare organizations, including Franciscan, are coming close to that 90% benchmark as they build a Kaizen culture.

Suggestion Systems Put the Burden on Managers

Another problem with suggestion systems is that the completed forms end up going to a manager and are added to the manager's long list of tasks. Since managers are typically very busy people, with other higher-priority items regularly popping up, a suggestion often gets buried in the pile.

Even when suggestions are reviewed and approved in a timely manner, the manager may assign it to someone other than the original submitter to review or implement. Is that person as passionate about the idea as the person who had the suggestion? Likely not. Both authors are grateful to Bodek for his clarity in teaching that suggestions are something *you* (pointing at another person) should do for me, while ideas are something that *I* can do for myself. With Kaizen, the person who identifies the problem works with the supervisor and other appropriate people to identify solutions and then typically leads the implementation of the best solution.

Winner Takes All Demoralizes the Rest

Additionally, some staff suggestion programs are framed as a "winner takes all" contest, where one idea is selected for implementation, with a prize awarded. Unfortunately, in a setting like that, all of the other great ideas are deemed losers, which is bound to be demotivating to those who did not have their valid ideas acted upon. With Kaizen, an organization can implement virtually every idea, large or small, and everybody can receive recognition for each of his or her improvements.

Suggestion Bonuses Cause More Trouble than They Are Worth

One other common dysfunction is the tendency of suggestion systems to pay out some percentage of any cost savings to employees. While this sounds good in principle, studies have shown that organizations that have larger payouts for suggestions actually get fewer suggestions.[14] It may seem counterintuitive, but suggestion payouts can hamper teamwork in a number of ways, including:

1. If the suggestion system only pays the person who originally had the idea, there's little incentive to work with others or there can be conflict about whose idea it was.
2. It can be incredibly time consuming to work to quantify the impact of a suggestion.
3. There can be a lot of controversy about how much a suggestion is really worth, especially if the payout is a percentage of the idea's value.
4. People will tend to focus only on large "home run" ideas or those that have a clearly quantifiable cost savings.
5. Payouts have often been based on merely having an idea as opposed to actually implementing anything that provided value or savings.

In the Kaizen approach, the ideal state would be to not pay staff for ideas, as we want to tap into people's natural intrinsic motivations. People feel pride when their ideas are listened to and when supervisors work together with them to drive change. In addition to just wanting to be heard, people are generally happy to make improvements that make their own work easier or provide better care for patients.

Franciscan does have some financial rewards, in addition to recognition that is given in different nonmonetary ways. The rewards at Franciscan are small, based on the documented implementation of an idea. The rewards are inclusive of all people who worked on a Kaizen team, not just the person with the original idea.

Kaizen and Lean

> The Toyota Production System—continual improvements toward profound evolution through full participation of all employees. There are no bounds to improvement. This is the basis for ongoing efforts by all employees to aim for kaizen (improvement), and refusing to ever be complacent.
>
> **—Sign in the Toyota Museum, Nagoya, Japan**

Hospitals like Franciscan around the world are using Kaizen, often as part of a broader Lean management initiative. The period of Lean healthcare adoption that started around 2000 has proven that healthcare organizations can improve when we have a highly engaged workforce focused on providing value to patients and minimizing waste in the delivery of care. Some of the most highly regarded Lean healthcare organizations include ThedaCare, Virginia Mason Medical Center (Seattle, Washington), Seattle Children's Hospital, Denver Health, Flinders Medical Centre (Australia), and the Royal Bolton Hospital NHS Foundation Trust (England).

Lean healthcare is a set of practices, a management system, and an organizational culture based on *Lean manufacturing* or the Toyota Production System.[15] It might seem strange for healthcare organizations to learn from a manufacturing company like Toyota. Rather than turning the *hospital* into a *factory*, Lean healthcare works tirelessly to provide ideal patient care—improving quality, reducing waiting times, and minimizing costs, all while furthering and enhancing the mission and caring nature of healthcare.

Kaizen: One of the Two Pillars of The Toyota Way and Lean

Formally published as an internal company document in 2001, The Toyota Way documented the two pillars of the automaker's management philosophy:

Continuous Improvement: Ongoing improvement in a structured and scientific way, in alignment with a long-term vision, by finding the root cause of problems and building consensus amongst a team.

Respect for People: Engender mutual trust and respect amongst all levels and stakeholders, challenging others to improve and maximize their abilities, encouraging personal and professional growth, and recognizing our inherent human limitations.

"Respect for people," at Toyota, applies broadly to employees, customers, suppliers, and their communities. Respect toward employees is a richer, more complex concept than just being nice to everybody. Respect means, for one, that people are constructively challenged to perform to the best of their abilities and to improve over time.

> Creating work that is meaningful and safe is the mark of profound respect for people.[16]
>
> **—John Toussaint, MD**
> *Former CEO, ThedaCare*

Respect for People = No Layoffs Due to Kaizen

One way that leading hospitals show respect for people is the recognition that they should never put employees in a position where their Kaizen ideas lead to layoffs. The American Hospital Association says a 2% Medicare cut, a loss of $41 billion over 8 years, would lead to 93,000 layoffs in 2013 and a total of 195,000 layoffs by 2021.[17] Kaizen and process improvement methodologies give hospitals an alternative to traditional cost-cutting approaches.

Park Nicollet Health Services in Minneapolis developed and instituted a "no layoff policy" early in their adoption of Lean management, says Steve Mattson, senior director of quality improvement at the health system. Matteson adds, "It was critical, as we began our cultural transformation and commitment, to focus on process problems, not people." The goal at Park Nicollet, he says, is to "bend the cost curve" by reducing waste (which leads to lower costs), adding that Lean is "a great way to involve teams in improving the work in their own areas."[18]

Kaizen and Respect Are Intertwined

Former Toyota leader Gary Convis summarized the connection by saying that an environment for continuous improvement can "only be created where there is respect for people."[19] These two mindsets, continuous improvement and respect for people, work together in a virtuous cycle and must be kept in balance.

> I think that [respect for people] is of profound importance because it means you are caring and you trust them to do the right thing.[20]
>
> **—Dr. Stephen R. Covey**

Organizations strive for continuous improvement out of respect for people, but it is that basic respect for people that helps make continuous improvement possible. If we do not truly listen and seek to understand the perspectives of all people in the workplace, regardless of their title or position, people will not collaborate constructively in improvement. If people feel ignored, disrespected, or devalued, they are likely to disengage from any attempts at improvement.

The way to ensure a sustained continuous improvement culture is to ensure each leader and each participant in the healthcare system practices respect for people—all stakeholders—each and every day.

Kaizen Closes Gaps between Staff and Leaders

In a primary care clinic, staff members complained that they spent too much time searching for the thermometer. Yes, *the* thermometer. Surprisingly, in a clinic with 5 or 6 nurses, 3 physicians, and 10 exam rooms, everybody was wasting time looking for the one and only digital thermometer.

This thermometer was supposed to be kept at the central nurses' station, which was inconvenient for all, because it meant walking from an exam room to get the thermometer when needed. It was a greater frustration, causing delay for patients, when the thermometer was inevitably being used by another nurse or medical assistant. In the course of brainstorming small improvements, the clinic team asked for a thermometer for each room, which would reduce the walking, waiting, and frustration for under $100 per room.

When senior leaders were given this proposal for spending approval, they expressed their shock and surprise. "What do you mean there is only one thermometer for that entire clinic?" asked one director. The clinic staff said they had never thought to ask for more thermometers, because they generally were never asked for their ideas on improving clinic operations, and they knew budgets were tight. The leadership team was not aware of the problem because they were rarely, if ever, present in the clinic and were unfamiliar with the details of the clinic's daily work.

Thanks to this discovery, the story did not end with only the approval of a $900 purchase request. This moment helped health system leaders realize that they also needed to find ways to continue closing the awareness gap to better support clinic staff and their patients over time.

Kaizen Values Creativity before Capital

There is a commonly used expression that says we should value "creativity over capital,"[21] as Kaizen thinking emphasizes finding simple, low-cost countermeasures and solutions. Shigeo Shingo, one of the creators of the Toyota Production System, chastised "catalog engineers" who simply bought solutions out of catalogs. In healthcare, there is a long-standing bias that the solution to problems automatically requires one or more of the following:

- More people
- More space
- More equipment

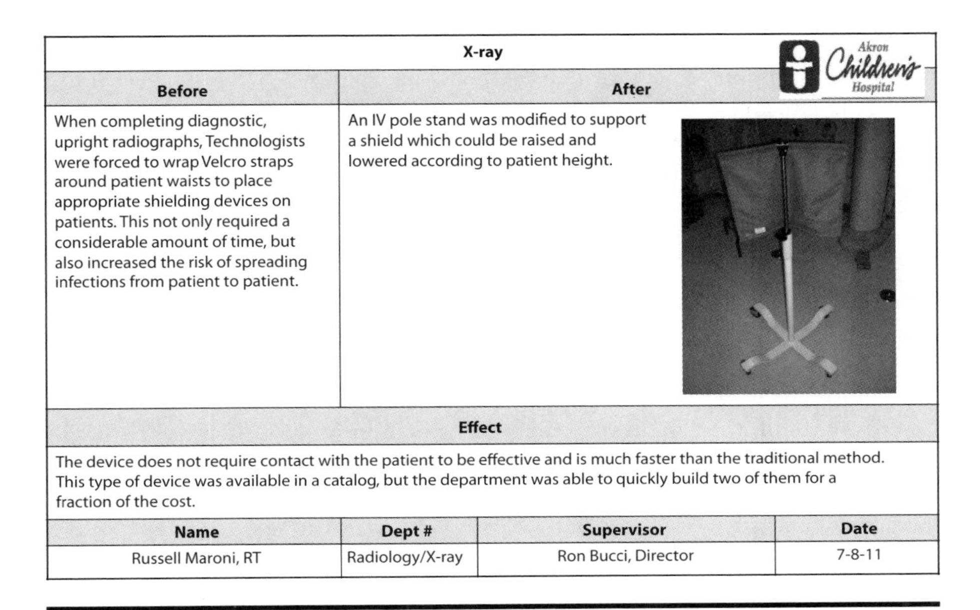

X-ray			
Before	**After**		
When completing diagnostic, upright radiographs, Technologists were forced to wrap Velcro straps around patient waists to place appropriate shielding devices on patients. This not only required a considerable amount of time, but also increased the risk of spreading infections from patient to patient.	An IV pole stand was modified to support a shield which could be raised and lowered according to patient height.		
Effect			
The device does not require contact with the patient to be effective and is much faster than the traditional method. This type of device was available in a catalog, but the department was able to quickly build two of them for a fraction of the cost.			
Name	**Dept #**	**Supervisor**	**Date**
Russell Maroni, RT	Radiology/X-ray	Ron Bucci, Director	7-8-11

Figure 2.2 A Kaizen that demonstrates the practice of "creativity over capital."

All of those options require more money. The Kaizen mindset does not mean we never spend money—sometimes, you do need more thermometers, for example, as the bare minimum requirement for effectively treating patients. We need to ensure people have what they need to provide proper patient care. But, we can challenge ourselves to first come up with creative solutions before spending money. Figure 2.2 is an example from Akron Children's Hospital (Akron, Ohio) where an X-ray technician created an adjustable patient shield from items they already had in the hospital.

In 2011, Masaaki Imai said, "If you have no money, use your brain … and if you have no brain, sweat it out!"[22] The vast majority of Kaizens at Franciscan are implemented for less than $100. People working on Kaizen are requested to use their creativity to find and test low-cost solutions before they resort to spending substantial sums of money.

Kaizen Helps Avoid Expensive Mistakes

At one hospital that Mark visited a few years back, the chief medical officer described how she had forced through the acceptance of a construction project to expand the number of emergency department exam rooms as an attempt to deal with overcrowding and having patients waiting in the hallways. Once the rooms were built, patient flow did not improve. At best, patients now had a room to wait in instead of a hallway. The executive admitted that her change (more rooms) did not lead to the improvement she desired (better patient flow). Hers was an example of rushing into an expensive, "big bang" solution. She considered it to be an expensive lesson learned, one she wished other hospitals would not repeat.

A Kaizen approach would have involved clinicians and staff members instead of being just a top-down executive decision. Through Kaizen, a hospital could experiment with a number of small-scale changes in a series of inexpensive and low-risk tests. A cross-functional team might learn that the root cause of emergency department waiting time was found far away—it is often the inpatient discharge process that is full of dysfunctions and delays. Or, the root cause could be the tradition of giving each patient an ER room during their visit. Some hospitals now move their "vertical" patients through an intake room, to a procedure room, then to an internal waiting area in order to increase throughput. Franciscan used this approach to reduce their ER *door-to-doctor* times and reduced patient length of stay by over 35%. Shorter length of stay means increased capacity.

Kaizen Reignites Our Inherent Creativity

> The greatest source of competitive advantage is not really cost or quality, but creativity.[23]
>
> **—John Micklethwait**
> *Editor-in-chief, The Economist*

Being creative simply means generating new ideas, and most people underestimate or discount their own creativity, thinking they can't contribute to Kaizen. People might not have big ideas, such as how to prevent all hospital-acquired infections, but they almost always have ideas about their own daily work, where they are the experts.

People, unfortunately, often have this creativity drummed out of them over time in the education system and the workplace. When employees are allowed and encouraged to creatively contribute to improve their own job performance and their organization in ways that are beyond their job role and scope, it might be surprising what can be accomplished.

When Norman Bodek teaches Quick and Easy Kaizen, he always asks audience members to raise their hands if they consider themselves to be a highly creative individual. A few hands sheepishly go up, maybe 2% of the room.[24] Norman explains that, in kindergarten, almost every child we know is highly creative and they demonstrate that constantly; however, after two years of school, they lose a shockingly large amount of their creative abilities.[25] Test taking teaches kids that it is important to memorize answers and conform; thoughts and ideas that do not fit the predetermined answers are wrong. The practices of the education system, including letter grades, which usually start in the first grade, create a culture of holding back one's creative abilities for the fear of being wrong.

> If there is no sense of trust in the organization, if people are preoccupied with protecting their backs ... creativity will be one of the first casualties.[26]
>
> **—Manfred F. R. Kets de Vries**
> *Clinical professor of leadership, INSEAD*

This fear of being wrong continues into the workplace, making people cautious to the point that they forget that they can be creative. With Kaizen, we want and need to reverse those learned thought processes and deeply engrained habits and barriers to creativity. Leaders need to build trust by participating directly and by not punishing "failed" attempts at Kaizen.

People Are the Ultimate Competitive Advantage

Toyota's website once stated: "Every Toyota team member is empowered with the ability to improve their work environment. This includes everything from quality and safety to the environment and productivity. Improvements and suggestions by team members are the cornerstone of Toyota's success."

A Toyota group leader from the Georgetown, Kentucky, factory said, "Toyota has long considered its ability to permanently resolve problems and improve stable processes as one of the company's competitive advantages. With an entire workforce charged with solving their workplace problems, the power of the intellectual capital of the company is tremendous."[27] While it is a cliché to say employees are your greatest asset, Toyota invests in people and their development because Toyota views its people as an *appreciating asset*,[28] while machines and buildings only depreciate over time.

In recent years, Toyota leaders have often referred to their approach as the Thinking Production System,[29] as they expect employees to have two jobs:

1. Do your work.
2. Improve your work.

In hospitals, staff members complain far too often that they are expected only to show up, keep their heads down, and do their jobs. By not engaging healthcare professionals in the improvement of their work, leaders waste a huge opportunity to improve patient care and the organization's bottom line. Hospitals can all hire the same architects, buy the same diagnostic equipment, and outfit operating rooms in the same way as some other leading hospital. Ultimately, however, healthcare performance is about people—not just their clinical skills, but also their participation in ongoing quality and process improvement.

Conclusion

The framework for a modern Kaizen culture was not an overnight invention. The roots of Kaizen thinking date back hundreds of years, if not longer. Today's healthcare organizations are learning from Toyota and other manufacturers that utilize the management practices of Lean and the Toyota Production System. Much as Toyota leaders build upon the teachings of those who came before them, healthcare organizations can, and should, allow their Kaizen philosophy to evolve, given other influences and inputs. Healthcare organizations have been talking about Kaizen for more than 20 years. Our hope is that healthcare is moving into a new era where Kaizen will be embraced more widely and more successfully.

Discussion Questions

- If you currently have suggestion boxes, are they being used effectively? What are the lessons learned?
- Does the word *Kaizen* cause problems or discomfort to people, to the point where you have to call it something else in your organization, like an *Idea System*?
- Does your staff know and practice a problem-solving methodology, like PDSA, that is based on the scientific method? Is it effectively getting to the root causes of problems?
- Can you think of a recent change that was not really an improvement? A change that was *resisted* by staff? What happened in those situations?
- How would you define *respect for people* in your organization? Is the idea widely practiced?
- Why is a no-layoffs approach important for Kaizen to be successful? Is that realistic for your organization? Why or why not?

Endnotes

1 Fujimoto, Takahiro, and Koichi Shimokawa, *The Birth of Lean: Conversations with Taiichi Ohno, Eiji Toyoda, and Other Figures Who Shaped Toyota Management* (Cambridge, MA: Lean Enterprise Institute, 2009), 42.
2 GBMP, *Thinking Outside the Suggestion Box: How to Create an Idea System That Works*, DVD, August, 2010.
3 ThedaCare Center for Healthare Value, *Thinking Lean at ThedaCare: Strategy Deployment*, DVD, 2011.
4 Jacobson, Gregory H., Nicole Streiff McCoin, Richard Lescallette, Stephan Russ, and Corey M. Slovis, "Kaizen: A Method of Process Improvement in the Emergency Department," *Academic Emergency Medicine* 16, no. 12 (2009): 1341–1349.
5 Maurer, Robert, *One Small Step Can Change Your Life: The Kaizen Way* (New York: Workman Publishing, 2004), 16.

6 Scholtes, Peter, Brian L. Joiner, and Barbara J. Steibel, *The Team Handbook* (Madison, WI: Oriel Incorporated, 2003), 1–7.
7 Graban, Mark, and Joseph E. Swartz, *Healthcare Kaizen: Engaging Front-Line Staff in Sustainable Continuous Improvements* (New York: Productivity Press, 2012), 335.
8 Zarbo, Richard, J. Mark Tuthill, et al., "The Henry Ford Production System: Reduction of Surgical Pathology In-Process Misidentification Defects by Bar Code–Specified Work Process Standardization," *American Journal of Clinical Pathology* 131 (2009): 468–477.
9 Rother, Mike, *Toyota Kata: Managing People for Improvement, Adaptiveness and Superior Results* (New York: McGraw-Hill, 2009), 139.
10 Robinson, Alan G., and Sam Stern, *Corporate Creativity: How Innovation and Improvement Actually Happen* (San Francisco: Berrett-Koehler, 1997), 72.
11 Graban, Mark, "Time & Motion Studies Are Not 'Discredited,' Just How They Are Used," LeanBlog.org, May 25, 2011, http://www.leanblog.org/2011/05/time-motion-studies-are-not-discredited-just-the-way-they-are-used/ (accessed August 27, 2011).
12 Seidelman, Ray, personal interview, July 2011.
13 GBMP, "Thinking Outside the Suggestion Box: How to Create an Idea System That Works," DVD, August 2010.
14 Robinson, Alan G., and Dean M. Schroeder, *Ideas Are Free: How the Idea Revolution Is Liberating People and Transforming Organizations* (San Francisco: Berrett-Koehler Publishers, 2006), 60.
15 Graban, Mark, *Lean Hospitals: Improving Quality, Patient Satisfaction, and Employee Engagement*, 2nd edition (New York: Productivity Press, 2011), 2.
16 Toussaint, John, "A Management, Leadership, and Board Road Map to Transforming Care for Patients." *Frontiers of Health Service Management*, March 2013, http://www.createvalue.org/data/blog/Frontiers_29_3%20Spring-FINAL_Toussaint%20exec%20summary.pdf (accessed May 5, 2013), 7.
17 Graban, Mark, "Lean as an Alternative to Mass Layoffs in Healthcare," *Becker's Hospital Review Online*, http://www.beckershospitalreview.com/hospital-management-administration/lean-as-an-alternative-to-mass-layoffs-in-healthcare.html (accessed May 6, 2013).
18 Graban, *Becker's*.
19 Dague, James, personal interview, July 2011.
20 Graban, Mark, "LeanBlog Podcast #91: Dr. Stephen Covey on Respect for People and Lean," http://www.leanblog.org/91 (Accessed January 26, 2013).
21 Kato, Isao, and Art Smalley, *Toyota Kaizen Methods* (New York: Productivity Press, 2010), 102.
22 Declaire, Joan, "Late to the Party: Confessions of a Lean-Hesitant Manager," DailyKaizen.org, July 8, 2011, http://dailykaizen.org/2011/07/08/ late-to-the-party-confessions-of-a-lean-hesitant-manager-by-joan-declaire/ (accessed May 6, 2013).
23 Micklethwait, John, and Adrian Woolridge, *The Right Nation: A Study in Conservatism in America* (New York: Penguin Press, 2004), 185.
24 Bodek, Norman, personal interview, August 2011.
25 Robinson, Ken, "Sir Ken Robinson: Do Schools Kill Creativity?" YouTube.com, 2006, http://youtu.be/iG9CE55wbtY (accessed January 25, 2013).
26 Gill, Roger, *Theory and Practice of Leadership*, (London: SAGE Publications, 2006), 141.
27 Rother, Mike, *Toyota Kata: Managing People for Improvement, Adaptiveness and Superior Results* (New York: McGraw-Hill, 2009), 14.

28 Liker, Jeffrey K., and Timothy N. Ogen, *Toyota under Fire: Lessons for Turning Crisis into Opportunity* (New York: McGraw-Hill, 2011), 33.
29 Public Affairs Division Toyota Motor Corporation, "The Thinking Production System," website, October 8, 2003, http://www.kellogg.northwestern.edu/course/opns430/modules/lean_operations/ThinkingToyota.pdf (accessed January 25, 2013).

Chapter 3

Types of Kaizen

Quick Take

- Kaizen can include large projects, weeklong events, and small improvements.
- Kaizens of all sizes and complexity follow the Plan-Do-Study-Act (PDSA) model.
- Rapid Improvement Events (or *Rapid Process Improvement Workshops*) are combined with *continuous daily improvement* at leading Lean health systems.
- Finding and implementing small improvements builds capabilities for innovation.
- Many innovations are discovered in the process of making small improvements.

> Such is the delicacy of man alone, that no object is produced to his liking. He finds that in everything there is need for improvement.
>
> **—Adam Smith**

The Continuous Improvement of a Lifesaving Innovation

One day in 2008, a man named Greg woke in the middle of the night with intense pain in his chest. Greg recalled, "Much like someone is standing on your chest, you know at that point you're having a heart attack." After arriving at Franciscan St. Francis Hospital, he remembers, "My whole body was shaking. I was thinking … I may not make it. I remember saying goodbye to my wife … and lots of tears … and pain." After receiving a percutaneous coronary intervention (PCI) well within the critical first 90 minutes, a procedure that opens an artery that feeds the heart muscle, he lived to talk about his experience. "My life was saved purely by the quick work of the heart physicians," he said.[1]

Greg benefited from a new protocol, EHART®, which was developed at Franciscan St. Francis Health. With this protocol, the average *door-to-balloon time* was reduced from 113 minutes to just 75 minutes. Patients receiving treatment within 90 minutes increased from 28% to 71%. Since "time is muscle," as some say,[2] the average heart attack size was reduced by 40%, and the average length of stay was reduced by two days[3] as the result of clearing the blockage sooner. Much of the improvement came from physician-driven systemic changes to the process, including activation of the catheterization lab by an emergency physician (instead of waiting for a cardiologist) and the immediate transfer of the patient to the cath lab by an in-house transfer team.[4]

The creation of the EHART protocol, a radical innovation (not formally driven by a Lean or Kaizen process), was not the endpoint for improvement. Rather, it was just the beginning. Thanks to *dozens* of Kaizen improvements, the health system not only sustained its improvements, but also continuously improved door-to-balloon times and patient care. Kaizen generally does not involve committees or slowly unfolding projects; these small Kaizens add up to a major impact for patients, caregivers, and our healthcare organizations.

How did Kaizens at Franciscan further improve heart attack patient care? In one example, Nathan Lowder, an ER nurse, developed a Kaizen that saved more than 20 minutes during the typical patient transfer from his ER to the cath lab by having all necessary medications given to the patient while being transported in the ambulance.

Additionally, Elizabeth Black, a nurse in the cath lab, noticed she was wasting valuable time during an EHART because she was always digging through a pile of different blood pressure cuffs. To reduce this waste, she quickly found some VELCRO® brand fasteners that she used to mount and organize the blood pressure cuffs of various sizes so they could be retrieved quickly. Later, maintenance ordered and installed more permanent holders, a Kaizen on top of the Kaizen.

Kaizen Means Continuous Improvement or Just Projects?

The word *Kaizen* is usually used in the context of small incremental changes to an existing process. Over the past 20 years, however, the word has also become associated with episodic workshops, often called *Kaizen Events*, which typically last for three to five days. That may seem puzzling to the reader. If a hospital department holds two or three weeklong events per year, how is that improvement the least bit continuous? Do these events really lead to sustained improvement?

If we use the translation of Kaizen that means "change for the better," then we can perhaps broaden its definition to include improvements of varying size, length, and complexity. Healthcare, at times, badly needs radical change such as the complete transformation of processes or the design of an entire building. But there are also great opportunities to make thousands of small improvements as well.

Three Levels of Kaizen

Toyota has taught us that there are different levels of Kaizen that are practiced in a high-performing organization, as illustrated in Figure 3.1. All three levels of Kaizen are based on the same Plan-Do-Study-Act (PDSA) problem-solving model and mindset.

Large Projects

Organizations often focus on a small number of large initiatives or projects, represented in the top portion of Figure 3.1, which might include building a new patient tower or implementing a new electronic medical records (EMR) system. Many of these major improvements are, even if managed well, high cost and

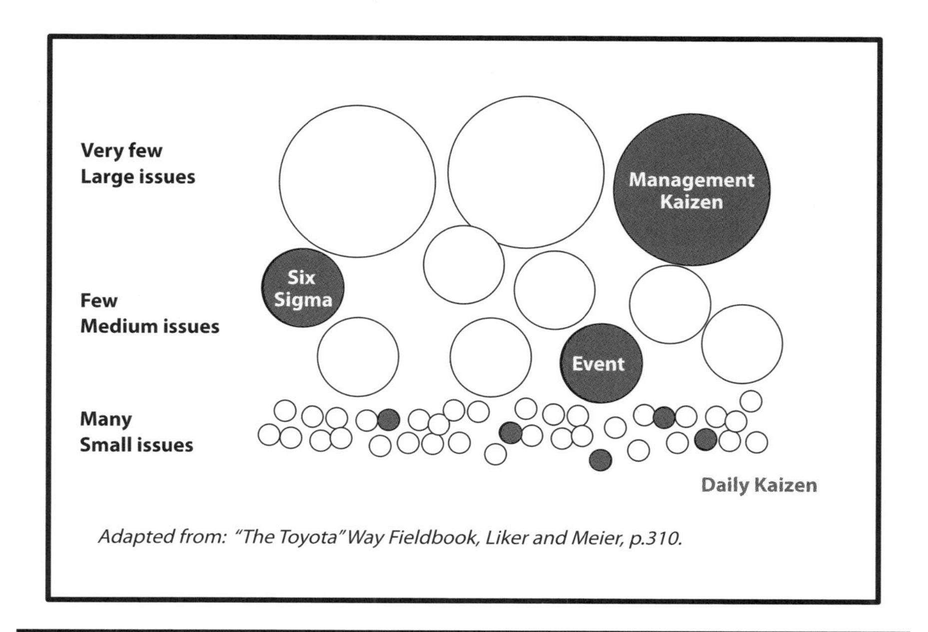

Figure 3.1 Illustration of the three levels of Kaizen improvements.

high risk—the opposite of small Kaizen improvements. Yet, these improvements are often very necessary. This level of improvement is often referred to as *management Kaizen* or *system Kaizen* in the Toyota framework. Major changes to our healthcare insurance and reimbursement system could be considered *system Kaizen*.

Mid-Sized Projects

The middle range of opportunities includes issues that impact the entirety of a department or a single patient value stream. These issues are less complex than management Kaizen, yet more complex than daily Kaizen. These are problems that might need to be solved by a cross-functional or cross-departmental team that is assembled for a small project. The methods used to address these opportunities can include weeklong events, Six Sigma projects, GE "work-outs,"[5] or A3 problem-solving reports.[6]

One example of a formal improvement event was conducted at Virginia Mason, which reduced operating room turnaround times from the baseline average of 30 minutes, allowing them to increase the utilization of existing rooms. The workshop team studied the existing work to find activities that were done in the room that could be safely performed outside the operating room. This allowed some prep or recovery steps to take place while another patient was in the room.

One activity that took time was the movement and positioning of patients from a gurney to the OR table. The team identified a type of bed that was versatile enough to be used for prep, procedure, and recovery, saving time and the need for staff to lift patients.[7] While that Kaizen required spending money, it was the type of case where the improved OR productivity more than paid for the beds.

After the Rapid Process Improvement Workshops (RPIWs), Virginia Mason Medical Center (VMMC) was able to perform 140 cases a week in four rooms, compared to 100 previously—a 40% improvement. Turnover time was reduced from more than 30 minutes to less than 15. Surgery prep time, from setup to suture, was reduced from 106 minutes to 85.[8]

Six Sigma is a quality improvement methodology that was created at Motorola in 1986[9] and was popularized by General Electric and their former CEO, Jack Welch, in the 1990s.[10] While Six Sigma is, like Lean, an approach for continuous quality improvement, Six Sigma is characterized by its focus on statistical analysis and the formal training of "belts" (such as lower-level Green Belts or more advanced Black Belts or Master Black Belts), people who lead defined improvement projects in an organization. By comparison, Kaizen methods can be used by everybody in an organization.

Smaller, Daily Improvements

Organizations have a certain capacity for large projects and events, but there is always more capacity for daily Kaizens that are managed through the Quick and Easy Kaizen process or Visual Idea Boards, as discussed more in Chapter 5. Examples of continuous daily improvement at ThedaCare include:

- Designating a standardized location for storing open patient care items in their room
- Finding an easier way to ambulate patients who need oxygen
- Creating a better list of medications for discharged patients to avoid confusion
- Figuring out how to have palliative care or hospice available on the weekend
- Teaching or reminding housekeeping where isolation signs are supposed to be stored[11]

Complementary Nature of the Levels of Kaizen

These three levels of Kaizen are complementary and supportive of each other. David Meier, a former group leader at Toyota's plant in Georgetown, Kentucky, teaches that he was involved in only a handful of formal improvement events during his ten years at Toyota. When Toyota did conduct events, the purpose was not short-term return on investment or even the improvement itself. According to Meier, the purpose of an event was for management to *learn* about Kaizen.[12] By learning about Kaizen, this same PDSA process could be applied to larger problems (system Kaizen) or smaller problems (daily Kaizen).

In The Toyota Way Fieldbook, Meier emphasized, "Many organizations fail to develop an effective process for capturing opportunities from all three categories. Quite often, the small category is overlooked entirely because these opportunities are viewed as 'insignificant' or offering 'not enough bang for the buck.' In addition, the medium and large opportunities are not fully exploited due to the small number of people being trained or qualified to resolve issues."[13]

At the Cancer Treatment Centers of America (CTCA), their hospitals started with daily continuous improvement, done through the framework of an A3 problem-solving model, a planning and problem-solving reporting methodology that is "core to the Toyota management system."[14] Herb DeBarba, vice president at CTCA, said they would "rather have 1000 little improvements than one big one." DeBarba adds, "There are a lot of healthcare institutions doing a great job with Kaizen weeks, but it's the grassroots effort that's making us most successful and allows us to help run better Kaizen events. We have very engaged front-line staff." DeBarba added, "When we pull the teams together to run 3- to 4-day events, we rarely have to do any training on the tools and principles. We can go right to work because they've practiced Kaizen on a small scale."[15]

Three Types of Kaizen at Children's Medical Center Dallas

The core laboratory at Children's Medical Center Dallas started its Lean journey in 2007 with an initial project led by coauthor Mark. The team, comprised of 6 laboratory professionals (medical technologists and laboratory assistants), studied their existing layout and process flows over the course of an initial 8-week phase. The team worked with their leaders and colleagues to design a new "core cell" layout that would represent a major overhaul of the lab's physical space and work processes.[16] This radical change in the layout represented a system Kaizen improvement effort.

While the lab expected that the system Kaizen efforts would benefit patients (faster turnaround times), staff (a better work environment), and the hospital (better financial performance), there was interest in changing to a Lean management system, embracing Lean and Kaizen as a way of thinking toward the broader goal of creating a *learning organization*.[17]

While they were working on the system Kaizen efforts, the laboratory's leaders and staff started working on small daily Kaizen improvements in their existing processes and existing space. Many of their examples and lessons learned are shared in Chapters 6 and 7 of *Healthcare Kaizen*.[18] The lab also conducted some improvement events using the GE "work-out" methodology,[19] covering the middle range of Kaizen.

The business case for the initial project was made based on the system Kaizen improvements. But daily Kaizen was incorporated from the start, which helped build confidence in the team that they could make meaningful changes at a larger level. Once the system Kaizen was completed and the new layout was in place, the culture of continuous improvement that had developed led laboratory staff members to make small daily Kaizen improvements of that new layout.

In many traditional organizations, a major change, such as a brand new layout, would be something that people dare not challenge or criticize. Yet, at Children's, the team and leaders embraced the PDSA mindset, allowing for small Kaizens to tweak the new process. The system Kaizen layout had been well thought out and had been simulated in a number of ways, so everybody had been confident that it would work. Yet, with nearly any major change, there were small details that could not be anticipated until the new layout was really in place. The study phase that was conducted after the Plan and Do phases allowed the team to adjust the layout and workflow in a way that provided even better care to patients. Jim Adams, senior director of laboratory operations commented, "Every operational improvement allows you to see new barriers that you couldn't see before, which is why we encourage and celebrate the ongoing improvement of our process."[20]

Events Are Powerful, but Not Enough

It is common in healthcare, as well as other industries, that the word *Kaizen* is always associated with the word *event*. These team-based improvement events, typically lasting three to five days, can be powerful, often resulting in dramatic changes and impressive results. Surely, weeklong improvement events have their place; however, some organizations get the impression that conducting these episodic events is the only path to improvement, as their consultant has perhaps taught them that the only way to learn Lean is by doing events. More mature Lean organizations have learned that they need to supplement events with other types of Kaizen, namely daily continuous improvement.

As Imai says, organizations cannot rely on intermittent projects. He said, in 2001, that doing a project and then "taking it easy for three months ... is not every day improvement."[21] In Imai's book *Kaizen*, the idea of weeklong events or workshops did not appear at all.

Mike Rother, in *Toyota Kata*, states it directly: "Projects and workshops ≠ continuous improvement,"[22] adding, "Relying on periodic improvements and innovations alone—only improving when we make a special effort or campaign—conceals a system that is static and vulnerable."[23] This vulnerable system might include the traditional top-down organizational culture that impedes daily improvement. Projects and events do not "require any particular managerial approach," says Rother, and "this may explain some of the popularity of workshops."[24]

The best Kaizen organizations, including those in healthcare, are improving each and every day, not only by changing the way frontline staff operate each day, but also changing the behaviors of leaders at all levels.

Basic Structure and Format of an Improvement Event

A classic improvement event starts on Monday, ending on Thursday or Friday. Some events have their work continue through Friday, while many, like those at ThedaCare, are four-day events that have a combined "report out" meeting on Friday where each team shares its event with an audience of about 200 people.[25]

An event is a team-based approach, with the ideal size of no more than 8 to 10 participants. Team members should represent a diversity of different roles, departments, and experience levels, including physicians and other clinicians. The team should include some outsiders (from another department or even a patient) who can ask good questions and provide a different perspective than those who are in the middle of the work every day, in addition to those who are experts in the current process. Events are often led by outside consultants or by facilitators from an internal department, sometimes referred to as a Kaizen Promotion Office.

A typical weeklong event might have the following schedule (adapted from *The Kaizen Event Planner*[26]):

Monday: Event kickoff, team training, observe and document current state process
Tuesday: Design future state, brainstorm improvement ideas, select and prioritize ideas
Wednesday: Design improvements, create standardized work, test the new process, get input from other stakeholders
Thursday: Continue design and test cycles, gain buy-in from others, finalize standardized work, prepare training materials
Friday: Train other workers, create sustainability plan, complete report, hold presentation and celebration

Events are generally action oriented, with the aim of making and testing improvements during the week. To make best use of the event time, the event week is actually part of a longer planning, execution, and follow-up cycle that might last 7 to 10 weeks. This time includes team selection, data collection, and other activities that help the event week run more smoothly, as well as postevent follow-ups to look at measuring improvement and updating standardized work. Many organizations measure results after the event at 30-, 60-, and 90-day intervals to ensure that the new process is being sustained.

Additional Challenges with Weeklong Events

Even with the success of events, there are some criticisms of this approach or, at least, an overreliance on events as a transformation strategy.

For one, the schedule usually says that a dedicated event team creates standardized work on Wednesday or Thursday. What about the people who were not working that day? Most hospital departments run 24 hours a day, 7 days a week, and there are many part-time employees who might not be working during an event week. It is difficult or impossible to get sufficient input from all staff and stakeholders in a single day. Postevent follow-up discussions might raise an issue that some staff are "resistant" to the new standardized work, but this is quite understandable if some staff did not have a say or did not even get trained properly on the new process. In some settings, such as a small primary care physician practice, a weeklong event could more easily involve everybody who works in that area. The event team can get input from others and gain agreement on the new process in meetings after the workday is over.

Another major challenge with events can be the lack of sustainability of improvement ideas generated during that week. Many organizations find that, after an event, people go back to the old way of working. You might ask, "Why would people go back to the old way of doing things if it was really better for patients and the organization?" Reverting to old methods could mean that a change was forced through, it was not really an improvement, people were not properly trained on the new process, or there is a lack of leadership to work through the difficulties of creating change in an organization.

Even Virginia Mason, with all of its success, has struggled to sustain improvements that came out of RPIWs. As reported in 2004, they were "only holding the gains on about 40% of those changes, partially because it is easy to slip back into old ways of doing things if there is a lack of accountability and follow-through."[27] By 2011, their leaders reported that 90% of projects showed sustained results after 90 days, but only 50% held results and methods 6 or 12 months out.[28]

Combining Different Types of Kaizen

One reason many in healthcare equate Kaizen with events is the success of organizations like ThedaCare and Virginia Mason Medical Center, both of which have used weeklong events and daily continuous improvement in their Lean journeys.

Virginia Mason Medical Center

Virginia Mason Medical Center has been recognized for using RPIWs as a cornerstone of its Virginia Mason Production System (VMPS).[29] Thanks to the improvements that resulted from its RPIWs, VMMC providers are able to spend more time with patients, leading to better care as well as higher patient and staff satisfaction.[30]

While receiving less publicity than the RPIWs, Virginia Mason has also used an approach to daily Kaizen called the *Everyday Lean Idea* (ELI) program. This program was started in 2005, while RPIWs were being started and before the formal VMPS name was adopted. The approaches work together; as Jennifer Phillips, innovation director said, "We have a range of improvement systems, but ELI is designed for small problems. It's not the only method and neither is RPIW." She added, "We're getting better at triaging problems to the right approach" as issues arise.[31]

Phillips says she describes the Everyday Lean Idea program as an approach for "workshop avoidance," as again, certain types of improvement do not require the full structure of an event or an RPIW, so "you just tackle it."[32] Because "staff see the problems every day,"[33] the ELI program is focused on the small-scale problems that people can address quickly in their own work area. One ELI improvement involved staff members taking a small action to improve communication about patients who had fallen previously or were generally a higher fall risk. A small laminated red star was added to the fall alert flags that were already being used outside of patient rooms. The idea worked and was spread to other floors after that initial testing.[34] Phillips added, "We're trying to create a culture where staff have permission and the capability to just fix things and make it better—to make it part of their everyday work."[35]

The scope of ELI improvements is intended to be "something within your control," said Phillips, adding that it "can be a challenge" when an improvement opportunity is outside of your control. VMMC sees some collaborative ELI efforts across departmental boundaries, but they are still working to improve communication about ideas that cross silos. "VMMC is working to create a strong culture of readily accepting feedback from another department and finding the time to work together on improvement," Phillips added.[36]

Using ELI with other Kaizen methods (including weeklong RPIWs), Kaizen Events (which are shorter in duration at VMMC), and "3P" exercises[37] (for designing physical spaces) allows VMMC to create a culture where improvement is both top-down and bottom-up.

ThedaCare

ThedaCare has been using Lean principles since 2003. Initially, Lean education and training was done exclusively through the use of weeklong Rapid Improvement Events (RIEs), leading to $27 million in savings over the first four years.[38] Through a series of RIEs over a number of years, ThedaCare reduced the door-to-balloon time for Code STEMI patients from an average of 92 minutes to just 37 minutes, and CyberKnife waiting time was reduced from 26 days to just 6, showing that Kaizen methods could not only reduce costs and improve quality, but could also improve access and reduce waiting times for patients.

ThedaCare's consultants initially insisted, "The only way to learn Lean is to get out to 'gemba' and do Kaizen [event] work."[39] ThedaCare's leaders realized, however, that with 5,500 employees in 40 sites, "getting each employee onto an RIE ... proved impossible even after seven years of nonstop Kaizen. Another method had to be found. The RIEs could not do all the teaching and training."[40] ThedaCare eventually moved beyond a sole reliance on weeklong events.

Gemba: A Japanese word meaning the "actual place," used for the place where value-adding work is done,[41] such as an exam room, the laboratory, an operating room, or the cafeteria.

To supplement the learning from RIEs, ThedaCare created Lean and Kaizen awareness training for all staff members. They have also created a leadership development curriculum and mentoring program that helps teach people to be *Lean leaders*. Kim Barnas, a senior vice president at ThedaCare, said, "The ultimate arrogance is to change the way people work, but not to change the way we manage."[42] ThedaCare's goal is to develop people and leaders to be able to solve problems and improve performance every day, not just through RIEs.

As ThedaCare's improvement system matured, it became more of an ongoing management system that focused on what they call *continuous daily improvement* (CDI). As of 2011, it is expected that 80% of their improvement comes from CDI activities.[43] ThedaCare reports that there were 3,663 documented improvements in 2010, with a goal to double that number in 2011.[44] ThedaCare ended 2011 with 9,500 improvements and had over 20,000 improvements in their hospitals in 2012.[45]

John Poole, senior vice president for the ThedaCare Improvement System, said CDI is "the hard work in the Lean world: culture change, day by day."[46] He reflected further on their evolution, saying:

> For several years, we were harvesting the low-hanging fruit, doing Kaizen Events and RIEs that took obvious waste out of value streams. It felt gut wrenching at the time, but in hindsight, it was the easy 20 percent of needed change. We hadn't yet created a culture of continuous improvements. We didn't have good development plans for those who would manage and supervise in the new environment, and we didn't know what the new environment would look like. We had the mechanics down, but we did not have a sustainable, reliable system. Now, the whole human side of Lean is unfolding before us.[47]

Dean Gruner, MD, the chief executive officer of ThedaCare, talked about staff members generating ideas for improvement and then helping categorize them into one of the following types of improvements, depending on how complicated they are, requiring either:

- Just a few hours of work
- More thought and a complete A3 (or, as they call it, a PDSA)
- A full Rapid Improvement Event cycle[48]

Avera McKennan

Some organizations have embraced Kaizen and have achieved great results from Lean management practices without relying on weeklong events or, in some cases, not using any at all.

Avera McKennan Hospital and University Health Center (Sioux Falls, South Dakota) is a 510-bed facility that is part of Avera Health. The hospital started its journey in 2004 in its laboratory, as many hospital labs in that period were early adopters of Lean methods.[49] Avera McKennan embraced a two-pronged approach of large multimonth transformation projects along with ongoing continuous improvement, including an approach they call *Bright Ideas*. All employees at Avera McKennan have received training in Lean methods, and their Excellence in Service and Process program is a major part of their organization's strategy. Avera McKennan has conducted Lean transformation improvements in other departments, including the emergency department, pharmacy, women's center, and inpatient units.[50]

Fred Slunecka, formerly the regional president of Avera McKennan Hospital and now the chief operating officer of Avera Health, argues that improvement events are most effective when used after a more thorough system Kaizen effort. Slunecka says a longer project is the only way "you can gather all the relevant data, test all the alternatives, and achieve the buy-in of a large group of employees" for major change. Slunecka argues that healthcare redesign is too complex to accomplish in a week, saying, "I believe that doing *just* Kaizen Events is like applying so many Band-Aids." Slunecka says he is aware of only one event that was attempted in the Avera McKennan pharmacy. He was unimpressed with the results, saying, "After a solid week of work by the brightest and most expensive minds in the building, the only change that was successfully maintained over a month's time was the relocation of a printer!"[51]

Kathy Maass, director of process excellence, whose team runs the longer Lean transformation projects internally, commented:

> Most health care processes are complex and the implementation methodology we use grants staff the time needed to incorporate safety, quality, and efficiency into the standard work.
>
> We have experimented some with the implementation methodology and have done some shorter projects. While this takes fewer resources from a busy department, employee engagement can be a challenge, since the front-line staff is less involved with data collection. The most successful projects are the ones where the front-line staff gathers the data and develops the standard work.[52]

Avera McKennan's experience shows how different levels of Kaizen fit together into a cohesive whole. Avera McKennan started with a *kaikaku* (radical change) improvement of their laboratory's layout and processes, followed by ongoing Kaizen improvements, said Leo Serrano, former director of laboratories at Avera McKennan. Serrano recounts how the lab started with their large system Kaizen project in 2004. This was followed by a number of 6- to 8-week projects, or group Kaizens in Imai's framework, in 2005, conducted with the full participation of staff. Serrano says the lab conducts a formal evaluation every two years as a small group Kaizen, with individual Kaizens being done continuously (using Bright Ideas), with team members implementing their ideas after being evaluated together with lab leadership. Serrano concludes, "We now have a group of Lean trained individuals, and we are all committed to maintaining the hard won gains. It is a cultural change, and now our staff do individual Kaizens without realizing that it is a Kaizen; it's simply the way we do things around here."[53]

Kaizen Leads to Innovation at Franciscan

Franciscan, to date, has not done many RIEs, but they plan on using that approach more often in the future in conjunction with other modes of Kaizen. Their Kaizen leaders wanted to build maturity and experience with daily Kaizen before taking on larger events. Franciscan's view is that methods like Six Sigma, Lean, RIEs, and Quick and Easy Kaizen all need to support and encourage continuous daily improvement and building a Kaizen culture as a way of life. Kaizen needs to be taught and integrated into everyone's thinking from the top level to the front line.

When people think of innovation, they think of major breakthroughs. Although many may not realize it, daily Kaizen is a form of incremental innovation—it is just that the innovations are usually small. Shuhei Iida, MD, the CEO of Nerima General Hospital (Tokyo, Japan), says, "If you keep doing Kaizen, you will get innovation." In Dr. Iida's experience, people cannot plan to innovate, but it happens by accident, when people discover large opportunities in the course of making smaller improvements. He says, "As you do Kaizen, you increase your chance of innovation, as you stumble into things. As you keep doing Kaizen, you also look for big jumps."[54]

> The best way to have a good idea is to have lots of ideas.
>
> **—Linus Pauling**
> *Scientist and Nobel Prize Winner*

Charles A. O'Reilly III, professor at Stanford University, and Michael L. Tushman, professor at Harvard Business School, studied 35 recent and significant breakthrough innovation attempts, and they conclude, "to flourish over the long run, most companies need to maintain a variety of innovation efforts."[55] They claim that for organizations to be most successful at innovation, they need to pursue not only breakthrough innovations, but also incremental innovations, which they define as small improvements in existing products and operations.

Studies at Procter & Gamble, Lego, and other organizations have shown that a distinguishing characteristic of highly innovative organizations is their willingness to try things and fail.[56] Highly innovative organizations are made up of people who are more willing to try something to learn and see if it will make them better. They look at failure as an important way to learn. They are also much faster at making adjustments to those things they try that are not working as well and much less likely to give up on improvement efforts. This is the PDSA cycle, and Kaizen, in practice.

At its core, Kaizen is about developing a culture of innovation, as the goals of large-scale innovation and continuous improvement are similar—to make something better. If Kaizen is used as a core of cultural transformation, it can develop a workforce that rethinks and improves the products and services that an organization provides in ways both large and small.

Franciscan COO Keith Jewell says, "Kaizen started the innovation journey for us. We were innovative 'in pockets.' Now we have 4,000 people whose job is to innovate every day—to reduce costs, and improve quality, productivity, safety, and satisfaction."[57]

Conclusion

Kaizen is proving to be an increasingly important part of Lean transformation efforts and healthcare improvement. When an organization takes a narrow view of Kaizen, such as equating it only with small improvements or only with weeklong events, success may be limited or unsustainable. We see that "change for the better" can come in different forms, ranging from major initiatives to daily continuous improvement. More importantly, these modes of improvement are compatible and mutually sustaining, because they are all based on the PDSA process and the scientific method. There are many different paths on the Kaizen journey, where some organizations start with smaller daily Kaizen improvements and then progress to larger changes, like the CTCA, while others have taken the opposite approach, such as Avera McKennan. Finally, as organizations like ThedaCare are demonstrating, these Kaizen activities should be tied to the strategy and goals of the broader organization, in addition to individuals making their own work easier.

Discussion Questions

- How would you determine if a problem needed to be solved with a large *system Kaizen* initiative, a weeklong event (RIE or RPIW), or a small *daily Kaizen* improvement?
- Has your organization faced any challenges related to weeklong improvement events or an overreliance on this improvement model?
- How do events help people better practice daily Kaizen? How can the practice of daily Kaizen lead to better events?
- For your organization, does it seem better to start with events and large initiatives, working toward daily continuous improvement, or to start with small changes, working up to events?

Endnotes

1 St. Francis Hospitals, "Greg's Story: Heart Attack Survival," video, http://youtu.be/kJf0_5Egl5g (January 25, 2013).

2 Toussaint, John, and Roger Gerard, *On the Mend: Revolutionizing Healthcare to Save Lives and Transform the Industry* (Cambridge, MA: Lean Enterprise Institute, 2010), 51.

3 Toussaint and Gerard, *On the Mend*, 51.

4 Khot, Umesh N., Michele L. Johnson, Curtis Ramsey, Monica B. Khot, Randall Todd, Saeed R. Shaikh, and William J. Berg, "Emergency Department Physician Activation of the Catheterization Laboratory and Immediate Transfer to an Immediately Available Catheterization Laboratory Reduce Door-to-Balloon Time in ST-Elevation Myocardial Infarction," *Circulation* 116 (2007): 67–76; originally published online June 11, 2007, http://circ.ahajournals.org/content/116/1/67.full.pdf?keytype=ref&ijkey=Js7FdI4iizHXh0N (accessed January 25, 2013).

5 GE, "What Is Six Sigma?" http://www.ge.com/sixsigma/SixSigma.pdf (accessed January 25, 2013).

6 Jimmerson, Cindy, *A3 Problem Solving for Healthcare* (New York: Productivity Press, 2007), 35.

7 Kenney, Charles, *Transforming Healthcare: Virginia Mason Medical Center's Pursuit of the Perfect Patient Experience* (New York: Productivity Press, 2010), 99.

8 Kenney, Transforming Healthcare, 101.

9 Motorola, "Six Things to Know About the History of Six Sigma," Motorola.com, 2004, https://mu.motorola.com/six_sigma_lessons/contemplate/assembler.asp?page=history_been_around (accessed January 25, 2013).

10 Process Quality Associates, "The Evolution of Six Sigma," website, http://www.pqa.net/ProdServices/sixsigma/W06002009.html (accessed January 25, 2013).

11 Graban, Mark, visit to ThedaCare, June 24, 2011.

12 Graban, Mark, "The Purpose of Kaizen Events," *LeanBlog.org*, November 5, 2009, http://www.leanblog.org/2009/11/purpose-of-kaizen-events/ (accessed January 25, 2013).

13 Liker and Meier, 309.

14 Shook, John, *Managing to Learn* (Cambridge, MA: Lean Enterprise Institute, 2008), vi.

15 DeBarba, Herb, personal interview, July 2011.

16 Adams, Jim, and Mark Graban, "CMCD's Lab Draws on Academics, Automakers, and Therapists to Realize Its Own Vision of Excellence," *Global Business and Organizational Excellence*, May/June 2011, 17, doi: 10.1002/joe.20383.

17 Adams and Graban, "CMCD's Lab," 17.

18 Graban, Mark and Joseph E. Swartz, *Healthcare Kaizen: Engaging Front-Line Staff in Sustainable Continuous Improvements* (New York: Productivity Press, 2012), 179.

19 Ulrich, Dave, Steve Kerr, and Ron Ashkenas, *The GE Work-Out* (New York: McGraw-Hill, 2002), 1.

20 Adams, Jim, personal interview, September 2011.

21 Imai, Masaaki, "Definition of Kaizen," YouTube, January 19, 2010. http://youtu.be/jRdTFis4-3Q (accessed January 25, 2013).

22 Rother, Mike, *Toyota Kata: Managing People for Improvement, Adaptiveness and Superior Results* (New York: McGraw-Hill, 2009), 11.

23 Rother, *Toyota Kata,* 11.

24 Rother, *Toyota Kata,* 26.

25 Toussaint, John, "John Shook Visits ThedaCare," website, http://www.createhealthcarevalue.com/blog/post/?bid=40 (accessed January 25, 2013).
26 Martin, Karen, and Mike Osterling, *The Kaizen Event Planner: Achieving Rapid Improvement in Office, Service, and Technical Environments* (New York: Productivity Press, 2007), 103.
27 St. Martin, Christina, "Seeking Perfection in Health Care: Applying the Toyota Production System to Medicine," *Performance and Practices of Successful Medical Groups: 2006 Report Based on 2005 Data* (Englewood, CO: Medical Group Management Association), 20.
28 Patterson, Sarah, paper presented at the Institute for Healthcare Improvement National Forum, December 7, 2011.
29 Shigujitsu USA, "Interview with Dr. Gary Kaplan, Virginia Mason Medical Center," website, http://www.shingijutsuusa.com/testimonials.html (accessed January 25, 2013).
30 Virginia Mason Medical Center, "VMPS Facts," website, 2010, https://www.virginiamason.org/workfiles/pdfdocs/press/vmps_fastfacts.pdf (accessed January 25, 2013).
31 Phillips, Jennifer, personal interview, July 2011.
32 Phillips, Jennifer, personal interview, July 2011.
33 Kenney, *Transforming Healthcare*, 162.
34 Kenney, *Transforming Healthcare,* 163.
35 Kenney, *Transforming Healthcare,* 163.
36 Phillips, Jennifer, personal interview, July 2011.
37 Grunden, Naida, *Lean-Led Hospital Design: Creating the Efficient Hospital of the Future* (New York: Productivity Press, 2012), 36.
38 Toussaint and Gerard, *On the Mend,* 3.
39 Toussaint and Gerard, *On the Mend,* 127.
40 Toussaint and Gerard, *On the Mend,* 128.
41 Marchwinski, Chet, and John Shook, *Lean Lexicon* (Brookline, MA: Lean Enterprise Institute, 2003), 23.
42 Barnas, Kim, "ThedaCare's Business Performance System: Sustaining Continuous Daily Improvement through Hospital Management in a Lean Environment," *The Joint Commission Journal on Quality and Patient Safety* 37, no. 9 (2011): 390.
43 ThedaCare Center for Healthcare Value, *Thinking Lean at ThedaCare: Strategy Deployment*, DVD (Appleton, WI: ThedaCare Center for Healthcare Value), 2011.
44 Barnas, "ThedaCare's Business Performance System," 390.
45 Toussaint, John, email exchange, January 2013.
46 Tonkin, Lea A.P., and Michael Bremer, "ThedaCare's Culture of Continuous Daily Improvements," *Target* 25, no. 1 (2009): 8.
47 Tonkin and Bremer, "ThedaCare's Culture," 7.
48 Graban, Mark, "Podcast #119: Dr. Dean Gruner, Strategy Deployment at ThedaCare," *LeanBlog.org,* May 31, 2011, http://www.leanblog.org/124 (accessed January 25, 2013).
49 Graban, Mark, *Lean Hospitals: Improving Quality, Patient Satisfaction, and Employee Engagement,* 2nd edition (New York: Productivity Press, 2011), 217.
50 Graban, *Lean Hospitals,* 218.
51 Slunecka, Fred, email correspondence, 2009.
52 Maass, Kathy, email correspondence, 2009.
53 Serrano, Leo, email correspondence, 2011.

54 Graban, Mark, "A Japanese Hospital CEO on Kaizen, Innovation & Breakthrough," http://www.leanblog.org/2012/11/a-japanese-hospital-ceo-on-kaizen-innovation-breakthrough/ (accessed January 25, 2013).
55 O'Reilly, Charles A., III, and Michael L. Tushman, "The Ambidextrous Organization," *Harvard Business Review* (April 2004): 74–81.
56 Lafley, A.G., and Ram Charan, *The Game-Changer: How You Can Drive Revenue and Profit Growth with Innovation* (New York, NY, Crown Business, 2008), 15.
57 Jewell, Keith, personal interview, December 2012.

Chapter 4

Creating a Kaizen Culture

Quick Take

- A Kaizen culture requires a large-scale organizational transformation.
- There might be many perceived barriers to Kaizen (such as lack of time), so those are some of the earliest problems to be solved.
- So-called resistance to change tends to go away when people initiate changes or are involved in the Kaizen process.
- Most people are actually highly creative and can participate in Kaizen.
- In a Kaizen culture, everything gets questioned and is open for improvement.
- Masaaki Imai teaches that early stages of a Kaizen culture focus on staff engagement and participation levels, not return on investment (ROI) or other performance targets.

Employee ideas are key to building a culture of high performance.

—Alan G. Robinson and Dean M. Schroeder[1]

Everyone Is Part of the Change Culture

Jenny Phelps, a nurse on a Franciscan nursing unit, noticed a breakdown in communication with food services personnel regarding whether or not a patient could have food. She decided to create a sign to signal when patients under her care could eat. After a brief chat with her supervisor, she presented the idea to the unit staff council, and they approved the idea. Jenny had the hospital print shop print and laminate some signs and tested them on her off day. Other nurses in the department liked them right away because the signs could remain at the doorway of the patient's room, being flipped over to change from "food allowed" to "NPO" (a Latin abbreviation for "nothing by mouth").

Within a week, the entire 36-bed nursing unit was using the new signs. The unit's manager, Christa Smiley, said, "In the past this wouldn't have happened so fast. It would have taken weeks and months and committee meetings, and convincing and selling. Now my nurses are getting used to seeing a good idea and quickly adopting the change. We now have a culture where improvement and change can happen faster and more effectively because everyone is part of the change culture."[2]

The Real Goal—Cultural Transformation

When coauthor Joe joined Franciscan as an employee in 2006, he asked Dr. Paul Strange, the vice president of quality services, what he envisioned as the goal for their Lean Six Sigma program. Dr. Strange said, "The real goal is a cultural transformation of our organization to one which constantly seeks to improve the quality of medical outcomes, make our processes safer, and to make our care more cost-effective."[3]

They set about understanding the organization's current state, what the future state needed to be, and how to create the required cultural change. In the past, improvement was the domain of certain personnel, such as performance improvement specialists. The new Lean Six Sigma program was poised to expand involvement much more widely, but it was not going to the entire organization anytime soon. A year into that program, there were two full-time Lean Six Sigma Black Belts and a few part-time Belts. The Six Sigma projects were 4 to 6 months in length and averaged 15 participants. The Lean events were shorter, but still required considerable preparation and follow-up time. Franciscan could only complete about 10 Lean or Six Sigma projects each year, and at that rate, it would take 26 years to involve all 4,000 employees.

The healthcare environment was getting more competitive. If Franciscan stayed on its current course, the future looked troubling, and leadership was concerned. Furthermore, changes were coming to healthcare faster and faster each year. They looked five years out and realized that the existing project-based Lean Six Sigma program could not grow fast enough to make the needed changes.

To make the long-term changes required at Franciscan, it was clear that everyone needed to be engaged in improvement. If a cultural transformation was needed, that meant a transformation of all employees. Franciscan's leaders realized that people must willingly transform themselves. So how do you get 4,000 people to want to do that? Kaizen was part of the answer.

Transformation starts by rethinking one's view of the world. Kaizen is more than something that is done—it is a way of thinking. Once the way of thinking permeates an organization and is practiced daily, it is the organization's culture. This chapter will explore what a culture of Kaizen looks like and how to start creating that culture.

Kaizen Grows Skills and Abilities

Franciscan has learned that Kaizen develops the skills and abilities of staff to solve problems and improve processes.

Bonnie Hicks, a staff nurse on Franciscan's postsurgical unit, found herself getting behind during the day, causing her to wait until shift's end to complete her documentation, which in turn would cause her to regularly get out of work late. She heard that Kaizen could help her save time and decided to get to the root cause of the delays, applying these methods to help her get home on time.

She asked her manager, "Can you explain this 'Kay-zen' thing that you are asking us to do?" To Bonnie, it seemed like extra work and something she didn't have time for—it was just one more thing that leadership was tasking her with. The unit manager carefully explained that it isn't extra work, "Kaizen has to be what we do each and every day—not what we do after we do our work. Our upfront investment today will save us time every day after today, and will make our job better, safer, and more enjoyable."

It wasn't until Bonnie completed a few Kaizens that she began understanding the benefit to her and her fellow nurses. Now, she has become one of the Kaizen leaders on her unit and carries a small notepad around with her to record the problems she sees and her ideas for solving them.

Bonnie has worked with her fellow nurses to implement dozens of Kaizens that have helped eliminate non-value-added activities that had frequently caused delays and frustrations on her unit. She was instrumental in moving numerous supplies closer to their point of use. Bonnie now has time to chart as she works and she is getting home on time more often. Kaizen has helped her do more of what she was called into her profession for—to connect with and care for others. Kaizen has also helped her develop her skills and abilities to make improvements, through her repeated practice of Kaizen, which when multiplied throughout a hospital system, is a vital part of creating a culture of continuous improvement.

Barriers to Kaizen

If we agree that individuals in healthcare want to do the right thing for patients and their coworkers, why is there so little Kaizen in some healthcare organizations? Or, why do we have so many improvements that are not sustained? What are the systemic barriers to Kaizen and how can they be reduced or eliminated?

Even the most highly educated healthcare professionals sometimes express skepticism about their ability to participate in Kaizen and continuous improvement on a daily basis. This can be especially true if their workplace has not engaged them in improvement, whether that has been over the entirety of a 30-year career or just the first three years of a young employee's working life. It can take time to gain confidence, which is why the Kaizen approach encourages people to start with small, local, low-cost, and low-risk improvements.

When we talk to leaders, physicians, and frontline staff, the idea of continuous improvement makes sense. It appeals to them rationally and intellectually, but people then start talking about barriers to Kaizen, which include the things listed in Table 4.1.

We have some tips for dealing with some of these barriers, whether real or perceived. But for many of these barriers, you will have to come up with your own countermeasures based on your organization. If you truly want to make Kaizen happen, then these barriers will be resolved through the creativity of your leaders and staff, applying continuous improvement and the Plan-Do-Study-Act (PDSA) cycle to the Kaizen process itself. As the radiology team at Children's Medical Center Dallas taught their colleagues, you have to shift your mindset from talking

Table 4.1 Barriers to Kaizen

Barriers to Kaizen
Others are resistant to change.
Lack of time.
"I get rewarded for firefighting."
Get in trouble for admitting we have problems.
"We forget to follow up on ideas."
"People don't think we need to change or improve."
People give up when not getting responses.
Lack of confidence in self or others.
Administration only wants cost savings (real or perceived).
Lack of trust in the organization.
Getting in trouble with coworkers.
"It's the other people who need to change."
Fear of losing jobs.
"I have people to do that for me; I delegate it."
"Perceived loss of control if I let employees do Kaizen."
"My people jump to solutions."

about "why this won't work" to reframing the discussion around "let's make this work."[4]

> Whether you think that you can, or that you can't, you are usually right.
>
> **—Henry Ford**

At Franciscan, coauthor Joe and other leaders have tried teaching Kaizen in the context of overcoming barriers, and it was not as effective as taking a more positive approach to making Kaizen successful. If you get a chance to take a race-driving course, the instructor will tell you that the car will automatically go where you are looking. If you look at an accident or look at the wall, you will be pulled into a crash. So, teaching Kaizen at Franciscan, the wall or the barriers are a reality, but they do not believe it should be the primary focus.

Ideally, we would not have to overcome barriers, as they should not be created in the first place. If you work with people the right way, you will not create as many barriers. This is the art of Kaizen. Just as there is an art to practicing medicine and nursing, there is also an art to practicing Kaizen, especially when it involves other people changing their behaviors or what they do. Of course, it is challenging to do it all right—even the best do not always get it right.

Resistance to Change

> Nothing so needs reforming as other people's habits.
>
> **—Mark Twain**
> *American author and humorist*

When people report that resistance to change is a problem in their organization, they can mean resistance to participating in a Kaizen program in general, or resistance to a specific Kaizen idea. When we feel or perceive resistance to a change, we should not blame the person or tell them simply, "don't be resistant, just get on board." Is there a legitimate problem that would be caused by the proposed change? Has there not been enough communication about the change and its expected impact? Did we not seek to get input from that person before announcing a change? Kaizen is more effective when we anticipate and try to proactively prevent resistance by engaging people in the right way, rather than just castigating them as being *change resistant.*

> 90% of resistance is cautionary.
>
> **—Shigeo Shingo**

Any perceived resistance should be the starting point for a discussion rather than a reason to stop talking and to stop working toward improvement.

Lack of Time—We're Too Busy

> People underestimate their capacity for change. There is never a right time to do a difficult thing. A leader's job is to help people have vision for their potential.
>
> —**John Arthur Porter**
> *Canadian sociologist*

It is not at all uncommon to get push back regarding Kaizen from staff and leadership, saying things like, "this is too much work," or "you're asking too much of us," or "we're too busy for this." It comes down to a core complaint of, "we don't have enough time."

Imai wrote that managers "should spend at least 50%" of their time on improvement.[5] How do we create time for Kaizen? Instead of accepting *lack of time* as an excuse that immediately ends the discussion about improvement, we should use our Lean problem-solving skills to understand why we have a lack of time. When we view lack of time as a problem to solve rather than an unsolvable excuse, we can introduce countermeasures that free up time for improvement. If we have too many meetings, we can take a fresh look at what meetings should even occur and which meetings are happening just because "it's always been that way." Since individual managers cannot arbitrarily decide to stop attending meetings, some organizations, including ThedaCare, have taken the approach of creating a two-hour *no-meeting zone* every morning.[6] During this window, the expectation is that managers will spend time attending team huddles in the gemba and coaching employees in their improvement efforts. The two hours freed up by the meeting-free zone is not intended to be used for answering email.

John Toussaint asks, "How can any of us be 'too busy' to be involved in delivering reliable care to our patients? I think the real reason this attitude exists is that most leaders do not want to change. The truth they do not utter is, 'Change is great as long as I don't have to do it.'"[7]

For staff members, we can focus initial Kaizens on freeing up time by eliminating waste. Paul O'Neill, former chair of the Pittsburgh Regional Health Initiative, said in an interview:

> We've found in the work that we've been doing that 50% of a nurse's time is spent doing things that don't add value, like looking for medications that aren't where they're supposed to be or looking for equipment that isn't where it's supposed to be. You want to identify and have the

> people in the process identify every aspect of waste every day, so that people can work on systems redesign to take out the waste.[8]

As people free up time by eliminating waste, some of that time can be dedicated to direct patient care or other value-adding activities. Some of the newly freed up time can also be used for additional Kaizen activity.

Oftentimes, when somebody has a Kaizen idea, there is not immediately time available to work on the Kaizen, or there is not time to get others involved for even a quick discussion. It is important to have a mechanism for holding ideas until we have time to work on them. Using slow times at the end of the day for Kaizen might be a more productive alternative, in the long run, to sending people home early when there is no direct work to be done.

> Evangeline Brock, a medical technologist at Children's Medical Center Dallas, says that they "take advantage of the slow times of year," which is the summer in the virology area, to spend more time on Kaizen improvements, especially ideas that might have accumulated on the Visual Idea Board during the year.

What a Kaizen Culture Feels Like

One might recognize a "Kaizen culture" when one sees it and experiences the unique and exciting feel of an organization where improvement is enthusiastically part of everybody's job. The following section describes some of the tangible characteristics of a culture of continuous improvement.

Everyone Is Engaged

Kaizen is about engagement. *Engagement* is a term that describes a combination of employee loyalty, commitment, and motivation. A recent report defines full engagement as "an alignment of maximum job satisfaction ('I like my work and do it well') with maximum job contribution ('I help achieve the goals of my organization')."[9] The report estimates that only one in three employees worldwide is engaged, with one in five being actively disengaged, meaning they are "disconnected from organizational priorities, often feel underutilized, and are clearly not getting what they need from work."[10] Employees become more strongly engaged when they contribute to making the organization better. Studies show that people want to be involved in decisions related to their work and workspace.

As a BlessingWhite study says, "Engaged employees are not just committed. They are not just passionate or proud. They have a line-of-sight on their own future and on the organization's mission and goals. They are enthused and in gear, using their talents and discretionary effort to make a difference in their employer's quest for sustainable business success."[11] A Kaizen culture has this level of engagement, personal mastery, and alignment.

According to an oft-cited 1946 study[12] that has been replicated more recently with similar results, the top two things employees want are appreciation and involvement. The Kaizen methodologies described in this book increase involvement and emphasize appreciation and recognition (which, again, does not necessarily mean financial recognition). Table 4.2 shows the top ten things that employees want in the workplace. The practice of Kaizen directly addresses most of those employee desires.

Drivers of Engagement

A 2005 Towers Perrin worldwide survey of more than 86,000 employees gives the top ten drivers of workplace engagement, as shown in Table 4.2.[13] Drivers 4 and 8 in the list indicate that employees want to be involved in decisions about their work. Drivers 1 and 2 indicate that employees value learning new skills. Kaizen enhances both learning and involvement, especially when an organization is focused on driver 5—customer and patient satisfaction.

Table 4.2 Top Ten Drivers of Workplace Engagement

Top Ten Drivers of Workplace Engagement
1. Opportunities to learn and develop new skills
2. Improved my skills and capabilities over the last year
3. Reputation of organization as a good employer
4. Input into decision making in my department
5. Organization focuses on customer satisfaction
6. Salary criteria are fair and consistent
7. Good collaboration across units
8. Appropriate amount of decision-making authority to do my job well
9. Senior management acts to ensure organization's long-term success
10. Senior management's interest in employees' well-being

Before this culture change took place at Franciscan, certain people were in charge of improvement, certain people were in charge of quality, certain people were in charge of finance, and everyone had their own siloed interests to look after. With a Kaizen culture, everyone is implementing improvements, and everyone is accountable and takes ownership for their part in making the healthcare organization extraordinary in multiple dimensions, including quality, patient safety, and cost. Masaaki Imai stresses that real culture change occurs when Kaizen is practiced daily by everyone in an organization, "from the CEO to the janitor."[14]

Franciscan COO Jewell said, "Employees want to make contributions, but sometimes they feel they need to be asked. I wanted them to know they are empowered to make change. I wanted them to know we want to give them the structure and infrastructure to make the necessary improvement to their work."[15]

Everyone Is Relentlessly Searching for Opportunities to Improve

Franciscan's culture also changed from one with a few people searching for improvement opportunities to one in which everyone is relentlessly looking for opportunities. One of the fundamental mindset differences that has been observed is that the most effective Kaizeneers were searching for improvement opportunities much more frequently than most. Kaizen permeated their lives as they were searching at work and, in many cases, at home.[16]

Before being exposed to Lean, people often confuse activity and motion with real value to the patient. *Value* is the activity that directly helps diagnose or treat a patient.[17] Patients ultimately value their health and quality of life, so one needs to look closely to see if activities contribute to that goal or not. If it is not adding value to the patient, it is waste.

People often gauge their own personal value in the workplace as "being busy," while Lean thinkers realize that much of the motion, searching, and workarounds in healthcare are actually waste, contributing no value to the patient and can, therefore, be eliminated by improving the process.

> Most people spend more time and energy going around problems than trying to solve them.
>
> **—Henry Ford**

After learning to distinguish waste from value, healthcare professionals start seeing waste where they never saw it before. Running around to search for a missing instrument is no longer unnoticed as a part of the everyday routine. It is now something that should be eliminated by making systemic improvements to prevent instruments from going missing. Once people start improving, they get swept up in the positive cycle of wanting to make more improvements, which means noticing and bringing forward more problems.

Laura Pettigrew, manager of medical records at Franciscan, said, "Staff are more aware of different ways of doing things because of being exposed to so many improvement ideas from the rest of the department and organization. They are now noticing opportunities for me and saying, 'that's a Kaizen.'"[18]

Nancy Mosier, former manager of pediatrics at Franciscan, said, "Kaizen supports the inverted pyramid" model of an organization chart, where staff members are shown at the top instead of their traditional place at the bottom. In Kaizen, our "staff searches for and finds improvement opportunities and works with their supervisor and manager to help them see the opportunities," she said. In Kaizen, staff members learn to enroll their peers and their leaders in their own ideas. Managers at all levels are involved in the model of "servant leadership."[19]

Patients and Families Are Happy

The purpose of business is to create and keep a customer.

—Peter Drucker

In a Kaizen culture, the primary customers, patients and their families, are happy with the services being provided because the value they want and need is delivered to them each time, exactly when needed. Furthermore, as the engagement studies earlier pointed out, a top-ten engagement driver is an organization that focuses on customer satisfaction. Employees define success as high-quality care and great service to patients and families. Employees realize that, if they can deliver better service to patients, they are contributing to revenue growth and the long-term strength of the organization, as well as their own job security. As Dr. Michel Tétreault, CEO of St. Boniface General Hospital, says, "If you want to satisfy patients, you must have engaged staff."[20]

Studies suggest that high employee satisfaction correlates with patient outcomes and lower rates of medical errors.[21] A Towers Watson study concluded, "It was found that employees' views of empowerment, career development opportunities, and teamwork influenced engagement. Further, employee engagement was a key predictor of patient satisfaction, leading to an increased likelihood that patients would recommend the network's hospitals to others."[22] It might seem reasonable to conclude that there is causation, not just correlation between these factors.

Mischelle Frank, manager of business transformation at Franciscan, led a team to redesign how food was delivered to patients in the postpartum nursing unit, with a focus on creating a "Wow Dining Experience." Her team turned in numerous Kaizens related to that redesign, such as several new recipes, how the food was presented, and the order in which the meals were prepared and delivered. These Kaizens had a positive effect on the patients' perceptions of the quality of the food in that unit, steadily rising from an average satisfaction score of 74% to 82% over 12 months.

Staff and Physicians Are Engaged

> Management's job is to create an environment where everybody may take joy in his work.[23]
>
> **—W. Edwards Deming, PhD**

Kaizen enables staff to reduce frustrations in their workspace. According to Jewell, "It is frustrating to have an inefficient system. Having a role in fixing their processes has a role in improving employee satisfaction. Encouraging employee engagement and involvement in running the hospital is making it better and that ability to contribute in a real way brings joy to the staff and into the workplace."[24] Physicians and surgeons can likewise participate in Kaizen, but are also happier as internal customers of improved processes that, for example, ensure procedures start on time more frequently, thanks to Kaizen.

Medical records has been the top-performing Kaizen department at Franciscan for three years in a row. When they started their Kaizen journey, their employee satisfaction was only at the 4th percentile in the nation. Three years into the program, their employee satisfaction had improved to the 63rd percentile.

Laura Pettigrew, their manager, credits the Kaizen program with being a significant reason for that improvement; as she said, "Staff used to rely on management to make all the decisions. Through Kaizen, we kept putting the onus on them to take ownership for their work and their work environment. We did a lot of coaching early on, and when they started to see that they could make better decisions than management, because they were closer to the work, and they started seeing the benefits, they were sold. Kaizen is how we gave our employees more power, control, and autonomy, and that translated into improved employee satisfaction."[25]

At Children's Medical Center Dallas, staff satisfaction increased in the core laboratory when measured 12 months into their use of Lean and Kaizen methods, as shown in Table 4.3. The satisfaction was measured on a scale of 1–5, with 5.0 being the highest. By 2013, about 5 years later, laboratory manager Clay York said, "We went from one of the worst in the division, for employee opinion scores, to one of the best. I think Lean and the changes we made took time to become part of the culture. I also think the Visual Idea Board (see Chapter 5) allowed us to listen to the staff, leading to improved scores. We have a much broader base providing ideas and they increasingly understand how they can communicate needs and their thoughts to leaders."[26]

In the Utah North Region of Intermountain Healthcare, they plan on continuing the annual staff surveys; the number of implemented staff ideas becomes a continuous measure of staff engagement, says Bart Sellers, regional manager of management engineering. Through this measure, they can gauge their progress toward their goal of every employee being a problem solver. "How better to engage employees than to have them involved in improvement?" asks Sellers.[27]

Table 4.3 Staff Satisfaction Scores 12 Months into a Lean Journey (out of 5.0)

	Before Lean	*12 Months after Starting*
I have the opportunity to do what I do best every day.	3.11	3.92
I feel free to make suggestions for improvement.	2.84	3.48
I feel secure in my job.	2.32	3.42
Stress at work is manageable.	2.43	3.23
I am satisfied with the lab as a place to work.	2.51	3.43
I would recommend my work area as a good place to work to others.	2.38	3.46
Grand Average of All Questions	2.96	3.69

The Workspace Is Clean, Orderly, and Safe

In a Lean environment, cleanliness, order, and safety go together. Lean methods like *5S*[28] are often misconstrued as being focused on just looking neat and tidy, but 5S is really about creating a safe, effective workplace where problems are readily visible. In this culture, problems occur less often because people are continuously working to prevent problems through better systems and processes. When problems do occur, they are investigated and resolved immediately or very quickly.

Ronda Frieje, pharmacy manager at Franciscan, noted, "Kaizen drove people to recognize the power over their own work environment to make change. Kaizen has driven positive changes in our department. Things just work better now. Our workspace is clean and orderly, and everything is where we've designed it to be. We separated look-alike, sound-alike, and different-strength drugs. I sleep better now knowing how much Kaizen effort we've put in over the last few years to ensure our patients are safe."[29]

In the MRI area of a hospital, a team in the early stages of a Kaizen program identified a situation that put patient safety at risk. Suction tubes and other safety supplies were not always restocked, leading to a situation where staff had to run around frantically, to other rooms, looking for supplies when a patient was vomiting in the MRI machine. Before Kaizen, staff would have just continued with their day after putting out the proverbial fire. In a Kaizen culture, staff and managers react to situations like this by improving processes in order to prevent future occurrences, working together in a blame-free manner. Better yet, people speak up and take action when potential risks are identified proactively.

Everyone Works Together

People in a Kaizen culture have accepted their expected role in workplace improvement. Supervisors and managers accept that they are no longer expected to have all the answers and their new role is to help employees shine. Improvement does not happen unless everyone works together to make it happen every day.

The pharmacy at Franciscan has experienced an impressive cultural change since starting its Kaizen journey. Ronda Freije, manager at the Indianapolis campus, said, "One of the biggest benefits Kaizen has brought to the pharmacy is a team culture. We work as a team now. Before, we tended to be split into cliques. The pharmacists would hang out with the pharmacists, the techs with the techs. Now, they are connected. Kaizen also developed our cross-shift teaming and cross-department teaming abilities. We work better with everyone now."[30]

Laura Pettigrew, manager of medical records at Franciscan, noticed a similar shift in collaboration, saying, "Our staff used to depend on leaders for day-to-day decision making on things like staffing and coverage. I knew, as a manager, I needed to get out of the day-to-day business so I could focus on developing the future of our department. I questioned what we could have our employees do. Now, if someone wants to take time off, it is up to them to work with others to get coverage. If they can't do it themselves, they come to me. Now, their communication is better." Laura said they now lead each other. "With Kaizen, they remind each other." Laura has noticed also that early on most Kaizens were solo, whereas now there are many more completed by small teams.[31]

Everything Gets Questioned

In a Kaizen culture, it is no longer acceptable to accept "Well, we've always done it that way" as an answer to the question of "Why do we do things this way?" Even practices that seem to be working well can be questioned, regardless of how big or how small the improvement is. We do not have to change everything, but nothing should be off limits from being questioned.

At ThedaCare, one contributor to the "door-to-balloon" time reduction was questioning why cardiologists had to consult and provide a second opinion if an emergency room physician saw signs of a "Code STEMI" event in the EKG printout. Waiting for the cardiologist would cause up to a 30-minute delay in that vital patient care. After allowing the emergency physicians to make the determination on their own, the next step was to question why a properly trained EMT could not make the determination in the ambulance.[32]

In another instance, Greg Kello, an environmental services supervisor for Franciscan, questioned why they were throwing away so many mop buckets. He discovered that they could clean and reuse those that had not been in isolation rooms. Many Kaizens come from asking the question, "Why have we always done it this way?"

For example, Kelly Butler, a lead nurse in the Franciscan NICU, questioned why a supply closet was locked. For years, nurses had to badge in to get into the closet to get supply items. Occasionally, a nurse would have their hands full, making it difficult to open the supply closet door. The lock was a problem that got in the way of ideal patient care, yet it went unnoticed. Once noticed, it was easy to get the lock removed, and as a result, nurses now just push open the closet door. A seasoned Kaizeneer may even question the purpose of the door.

Small Successes Lead to Bigger Successes

If we start with the small improvements first, we will then be more likely to adopt the habit of continuous improvement in all aspects of our life. With practice, everyone will be ready to tackle more aggressive and substantial Kaizens. Small changes are easier to implement and sustain. Making small changes is also more fun and satisfying than trying to make one large change.

Starting small with Kaizen has these attributes:

- They are small changes.
- They are within an employee's span of control.
- They are simple ideas.
- They do not negatively impact other departments or patients.

As people get better at making small improvements, they naturally try larger improvements. Over time, staff members become more comfortable with change as it becomes a way of life. Staff learn new skills, become more highly developed in problem solving, and gain confidence in making changes. Practicing with small changes makes the larger changes seem more possible.

When Kaizen was initiated at St. Francis, nearly all the Kaizens turned in were small improvements. Four years into the program, about 20% of the Kaizens turned in could be classified as medium to large improvements.

> Yard by yard, life is hard. Inch by inch, life's a cinch.
>
> **—Anonymous**

Nancy Mosier, former manager of pediatrics at Franciscan, said, "It is important to note that Kaizen is not just small changes for the better—it can be big improvements. However, we want our staff to start small and take baby steps so they can develop their skills and abilities over time to do Kaizen."[33]

> We can do no great things, only small things with great love.
>
> **—Mother Teresa**

You know the old saying about how to eat an elephant—one bite at a time. The same holds true for Kaizen. Any size project can be more easily completed if it is broken down into a series of small tasks. At the same time, one needs to be cautious that focusing on small tasks does not lead to losing sight of the overall process or needs of patients.

Imai's Three Stages of a Kaizen Culture

It is easy to say that you want to have a Kaizen culture in your organization. Changing an organization's culture may take a year in a single department, but it can take many years or a decade to change the culture of a larger organization. Starting small with baby steps can lead to larger improvements and culture change. Masaaki Imai shares a framework that includes three separate stages that one might expect to go through in a Kaizen journey.

In the first stage, supervisors must make every effort to help employees bring forward improvement ideas, no matter how small. The initial focus is on improving one's own job and area. Leaders need to say yes to virtually every idea in order to build interest and enthusiasm for Kaizen.

In the second stage, which might be a year or two into this journey, the organization starts to provide more education for staff on problem-solving skills, so people can provide better quality suggestions.

Imai emphasizes that "only in the third stage" should managers be concerned about the financial payback or ROI of improvements, once employees are interested and educated about Kaizen. He writes that organizations tend to struggle with Kaizen when they want to jump immediately to the third stage instead of thinking of the development of a Kaizen culture over a "five to ten-year span."[34]

Conclusion

Kaizen systematically creates an environment where change becomes the normal way of life. This results in a culture where people embrace change instead of being mislabeled as "resistant." Kaizen leads to greater adoption of change, of all types, as small changes lead to bigger changes. A Kaizen culture becomes a culture that is more responsive to the changing demands and landscape of healthcare. Healthcare professionals in a Kaizen culture are not just happy; they find more meaningful engagement in their important work.

The role of leaders is critical in creating a Kaizen culture. The behaviors and mindsets that are required, at all levels, to encourage Kaizen are described more fully in Chapters 6 and 7 and throughout the book. Transforming a culture requires that we advance beyond the status quo by driving toward the possibilities of perfection.

Discussion Questions

- How engaged are the people in your workplace? Other than surveys, how can you gauge overall engagement? How can you tell if an individual is highly engaged?
- What are the most commonly cited barriers to Kaizen in your organization? What steps can you take to counter them?
- How can you free up more time for your own Kaizen participation? To free up time for others in your organization?
- Why might it be important to not focus too much on cost savings in the early stages of a Kaizen program?
- Are there special actions that we need to take to engage our physicians in Kaizen?
- How do you know when your organization's culture has changed?

Endnotes

1 Robinson, Alan G., and Dean M. Schroeder, *Ideas Are Free: How the Idea Revolution Is Liberating People and Transforming Organizations* (San Francisco: Berrett-Koehler Publishers, 2006), 262.
2 Smiley, Christa, personal interview, March 2011.
3 Strange, Paul, MD, personal interview, December 2012.
4 Graban, Mark, and Joseph E. Swartz, *Healthcare Kaizen: Engaging Front-Line Staff in Sustainable Continuous Improvements* (New York: Productivity Press, 2012), 145.
5 Imai, Masaaki, *KAIZEN: The Key to Japan's Competitive Success* (New York: McGraw-Hill, 1986), 83.
6 Barnas, Kim, "ThedaCare's Business Performance System: Sustaining Continuous Daily Improvement through Hospital Management in a Lean Environment," *The Joint Commission Journal on Quality and Patient Safety* 37, no. 9 (2011): 393.
7 Tousaint, John, "A Management, Leadership, and Board Road Map to Transforming Care for Patients," *Frontiers of Health Service Management,* March 2013, http://www.createvalue.org/data/blog/Frontiers-29-3%20spring-FINAL-Toussaint%20exec%20summary.pdf (accessed May 5, 2013), 10.
8 Graban, Mark, "Quotes from the Paul O'Neill Podcast Interview on Patient Safety," LeanBlog.org, 2011, http://www.leanblog.org/2011/07/text-and-quotes-from-the-paul-oneill-podcast-interview/ (accessed January 25, 2013).
9 Blessing White, "Employee Engagement Report 2011," January 2011, http://www.blessingwhite.com/eee-report.asp (accessed May 5, 2013).
10 Blessing White, 6.
11 Blessing White, 5.
12 Labor Relations Institute of New York, "Survey: Foreman Facts," 1946, http://contentnetwork.org/business/1563.html (accessed January 25, 2013).
13 Towers Perrin HR Services, "Winning Strategies for a Global Workforce: Attracting, Retaining and Engaging Employees for Competitive Advantage," 2006, http://www.towersperrin.com/tp/getwebcachedoc?webc=HRS/USA/2006/200602/GWS.pdf.

14 Imai, Masaaki, "Definition of KAIZEN," http://youtu.be/jRdTFis4-3Q.
15 Jewell, Keith, personal interview, December 2012.
16 Graban and Swartz, *Healthcare Kaizen*, 337.
17 Graban, Mark, *Lean Hospitals: Improving Quality, Patient Safety, and Employee Engagement,* 2nd Edition (New York, Productivity Press, 2012), 32.
18 Pettigrew, Laura, personal interview, August 2011.
19 Greenleaf, Robert K., and Larry C. Spears, *Servant Leadership: A Journey into the Nature of Legitimate Power and Greatness,* 25th Anniversary Edition (New York: Paulist Press, 2002), 21.
20 Tétreault, Michel, MD, personal interview, January 2013.
21 Rathert, C., and D.R. May, "Health Care Work Environments, Employee Satisfaction, and Patient Safety: Care Provider Perspectives," *Health Care Management Review* 32, no. 1 (2007): 2–11.
22 Towers Watson, "Committed to Health: A Large Hospital Network Links Employee Engagement With Patient Satisfaction to Maximize Competitive Strength," Case Study, 2010, http://www.towerswatson.com/assets/pdf/1549/Healthcare_Case-Study_4-12.pdf (accessed January 25, 2013).
23 Neave, Harry R., *The Deming Dimension* (Knoxville: SPC Press, 1990), 198.
24 Jewell, Keith, personal interview, December 2012.
25 Pettigrew, Laura, personal interview, August 2011.
26 York, Clay, personal interview, January 2013.
27 Sellers, Bart, personal interview, September 2011.
28 Graban, *Lean Hospitals*, 89.
29 Frieje, Ronda, personal interview, June 2011.
30 Frieje, Ronda, personal interview, June 2011.
31 Pettigrew, Laura, personal interview, August 2011.
32 Toussaint, John, and Roger Gerard, *On the Mend: Revolutionizing Healthcare to Save Lives and Transform the Industry* (Cambridge, MA: Lean Enterprise Institute, 2010), 43.
33 Moiser, Nancy, personal interview, July 2011.
34 Imai, Masaaki, *KAIZEN: The Key to Japan's Competitive Success* (New York: McGraw-Hill, 1986), 113.

Chapter 5

Daily Kaizen Methods

Quick Take

- Kaizen is a practice best learned by doing.
- The five steps of the Kaizen process are to find, discuss, implement, document, and share improvement ideas.
- *Quick and Easy* Kaizen ideas are those that are quick and easy to test, implement, and review.
- Kaizen ideas can come from employees who are in their first week of work, if the culture encourages and allows it.
- *Visual Idea Boards* help make the entire improvement process visible and transparent to all who are involved—a major improvement over traditional suggestion boxes.
- Electronic systems for managing Kaizen are being used more frequently in healthcare.

You can't learn how to ride a bike by reading a book on physics.

—Unknown

Fresh Eyes Can See Waste That Hid before Your Eyes

During her new employee orientation at Children's Medical Center, a new microbiology technologist, Lisa, received more than three hours of introductory training about Lean and Kaizen. In the session, Lisa was told that her ideas were welcome and that she should expect to hold her leaders accountable for maintaining that environment. Joy Eckhardt, quality and regulatory compliance manager, says, "New people are often cautious to speak up,"[1] so it is important to teach this continuous improvement culture from day one.

In her very first week in the lab, Lisa had an idea for making her work easier. She noticed it was difficult to reach over the computer to get the Gram stain controls. As taught in the orientation, Lisa wrote her improvement idea on a card that she placed on a specified bulletin board for all to see. Unlike some workplaces, Lisa was not ignored because she was new, nor was she told to be quiet and just do her job; her voice was heard. The improvement was made and completed the very next day.

Evangeline Brock, a senior technologist in virology, commented that it is very common for new employees to discover waste and problems that go unseen by those who have worked in an area for years. Instead of being satisfied with the way things have always been, the microbiology department demonstrates that they value everybody's ideas. This was a small improvement, but Lisa's enthusiasm for Kaizen, for Lean, and for her workplace is very apparent when you talk to her. One can hope that this small idea in a new employee's first week of work will lead to a career full of Kaizen.

After starting with small ideas, such as moving things, employees usually move to "ideas with greater impact," says Eckhardt. Bernice Garner, the manager of the microbiology lab, noticed that after starting with "small, easier things"[2] in initial suggestions, the team eventually gained the confidence to later propose a new department layout to her, which was delivered to her, complete with drawings. In the main lab, Eckhardt recalled how the third shift team took the initiative to move some instruments, saying, "Before Lean, people wouldn't have even thought that the layout could be better; they would have just accepted it, as is."

It took time, Garner said, for people to learn that they "owned their ideas" and that the manager is "part of the team" instead of being the only person who could make or approve suggestions. Garner emphasizes that ideas are discussed at team huddles, and it is "more a matter of team consensus" to say yes or no to an idea. Her role, as manager, is to highlight regulatory concerns or other issues that her employees might not know about. The method used at Children's Medical Center Dallas, the *Visual Idea Board*, described later in this chapter, has been instrumental in building a Kaizen culture.

The 5 Steps of Kaizen

> Quick and Easy Kaizen simply asks you to stop for a moment and look for opportunities to improve.[3]
>
> **—Bunji Tozawa and Norman Bodek**

Norman Bodek is an American author who has played a critical role in the further spread of Kaizen practices and philosophies from Japan to the rest of the world. As coauthor of the 2001 book *The Idea Generator: Quick and Easy Kaizen*, Bodek shared a method for employees to document small changes they could make to their own work.

The Kaizen program at Franciscan is based upon this method, with its emphasis on documenting and sharing all of the small staff-driven improvements that are made in the workplace. *Quick and Easy Kaizen* emphasizes that we want (and often need) to start with Kaizens that are simple, or those that are quick and easy to test, implement, and review. If we start with the simple ones, we will then be more likely to adopt the habit of continuous improvement in all aspects of our life, as we learn about Kaizen and build confidence in our ability to improve. Over time, we will be able to tackle more aggressive and substantial improvements, such as Kaizen events or larger improvement projects.

There are five basic steps to Kaizen:

1. **Find:** Search for opportunities for improvement or problems to solve.
2. **Discuss:** Discuss the idea with your team and your supervisor.
3. **Implement:** A change for the better must be implemented to be a Kaizen.
4. **Document:** Document the Kaizen by creating a simple report called a Kaizen Report.
5. **Share:** Post it, review it, and discuss.

These five steps are simple and easy to learn and teach. Furthermore, they fit well in the healthcare environment. As Susan McRoberts, chief nursing officer at Franciscan, said, "I am surprised at how well Kaizen has integrated into everything we do. Our nurses are applying Kaizen to their daily routine, and now they are finding more opportunities, discussing more ideas with their colleagues, implementing more improvements, documenting more of those improvements, and sharing more than they ever have. Kaizen has become an integral part of our nursing culture."[4]

> At first I had difficulty finding Kaizen opportunities, but over time, as I practiced, it became easier and easier for me to see them.
>
> **—Melissa Horne**
> *Pharmacy technician, Franciscan St. Francis Health*
> *Completed 109 Kaizens in 2010*

Step 1: Find

The first of the five steps of Kaizen is to search for and find an opportunity for improvement or a problem to resolve.

> **Problem, Opportunity, or Both:** Some people prefer to define every problem as an opportunity, because the word *problem* often has negative connotations like doubt, difficulty, and uncertainty. However, Franciscan does not shy away from the word problem.

Kaizen often starts by focusing on the frustrations each staff member or leader experiences each day. Kaizen builds on one's own intrinsic motivation to make improvements for one's own benefit.

Staff can start generating Kaizens by searching for ideas that make their work any or all of the following:

- Easier
- Safer
- More interesting

Or they can look for any improvements that would build skills, capabilities, and knowledge.

For example, two nurses in Franciscan's endoscopy department rethought the bronchoscopy kits. It turned out that 65% of the items in the kit were not needed, so the kit was streamlined, resulting in a small cost savings and a very large reduction in nonbiodegradable plastics, as documented in Figure 5.1.

This Kaizen was forwarded by managers to staff across the entire Franciscan system to show the power of small, staff-driven improvements. This Kaizen went beyond "win–win" by benefitting:

- **Staff**—by saving time with the handling of fewer items
- **Patients**—by simplifying the items required for a procedure, minimizing the probability of a delay
- **Organization**—by saving costs and by having happier staff and patients
- **Society**—by reducing plastic waste

Beech Grove Endoscopy Waste Savings — Franciscan ST. FRANCIS HEALTH

Before	After
Bronchoscopy kits contained supplies that were frequently not used and were discarded.	The staff of the Endoscopy unit in Beech Grove reviewed the contents of pre-packaged bronchoscopy kits we receive from a local company.

The Effect

We were able to eliminate approximately 65% of the products, including a very large amount of plastic. Therefore, in addition to some cost savings, we were able to make significant reduction in non-biodegradeable waste.

Name	Supervisor	Date	Savings
Jayne Harris, RN, BSN Toni Perkins, RN	Loretta Hall	3/20/08	65% less waste. ~$300/year

Figure 5.1 Example of a win–win Kaizen.

Start Small

At Franciscan, the confidence to implement Kaizens did not just suddenly appear. It took several years of practice, trial and error, and coaching for our stronger Kaizeneers to grow their Kaizen skills and to gain enough confidence to become really proficient at Kaizen. Nancy Mosier, manager of Pediatrics at Franciscan, said, "I start where they are. If they are a beginner to Kaizen, I encourage them to take baby steps and start small. I approve their Kaizens right away to get wins, especially[5] if their Kaizen only impacts them. I want them to gain confidence in the process." As people gain confidence in their abilities, leaders can challenge them and push them to improve their problem-solving skills.

> It's lack of faith that makes people afraid of meeting challenges, and I believed in myself.
>
> —**Muhammad Ali**
> *World champion boxer*

Laura Pettigrew, manager of medical records at Franciscan, said, "Now staff members are taking more initiative because they have more self-confidence that they can take control over improving their job duties."[6]

Step 2: Discuss

After identifying a problem or opportunity, a Kaizeneer then discusses the idea with his or her team and immediate supervisor. In Kaizen, leaders certainly encourage employees to generate and take action on ideas, but that does not mean that supervisors or other managers are left completely out of the loop. Kaizen is not slow and bureaucratic, like a traditional suggestion box, nor is it a "free-for-all" where everybody just takes actions in isolation from their teammates and leaders.

Individuals are creative, but ideas tend to be better and more willingly adopted when we involve other people.

In this phase of the process, a person with an idea should:

- Discuss the idea with their direct supervisor.
- Discuss the opportunity, problem, or idea with team members.
- Quantify the expected benefits of the idea, where possible.

Say "Yes"

The supervisor's job is to discover how to say "yes" to the Kaizen. As much as possible, supervisors should approve Kaizens. In one of Masaaki Imai's benchmark companies, Aisin-Warner, 99% of ideas were accepted and implemented.[7]

In Franciscan's pharmacy, where every Kaizen is strictly scrutinized for any possible patient safety ramifications, more than 89% of Kaizens are implemented in some form. Unlike the old suggestion box process, in which managers and leaders voted "yes" or "no" in a very disconnected way, in cases in which the proposed change cannot be implemented, the supervisor helps the Kaizeneer modify her Kaizen into something workable. Kaizen is about the dialogue between employees and their supervisors, working together to find something that can be done to solve their problem.

Many Kaizens can be easily undone or reversed if the change does not turn out to be a demonstrated improvement. Again, as long as it's not anticipated that safety or quality will be harmed, the tendency should be to say "yes, let's try this idea." Following the Plan-Do-Study-Act (PDSA) cycle, we can test the idea and decide if we should stick with the change, go back to the old method, or try something altogether different.

If an idea involves spending a lot of money, then supervisors may need to work with the Kaizeneer to find a more creative or less expensive approach that addresses the underlying problem or opportunity. Although it's not critical to have a large return on investment (ROI) for each Kaizen, supervisors and managers do need to be cognizant of the short-term financial and cultural realities in their organization, meaning that Kaizen cannot be a "blank check" for staff members to spend, keeping in mind the Kaizen goal of "creativity before capital."

Lynne Meredith, director of revenue management at Franciscan, coaches her staff on Kaizens that seem unworkable or impractical by saying, "That is a great idea, but let's take that one step further," and she helps them think through a better solution.

At Franciscan we discovered that, if you disallow one Kaizen idea, you may turn that person off from the process altogether in early stages. People can get discouraged when Kaizens get blocked or delayed. If enough Kaizens are turned down, the Kaizen program can die, as happened with so many older suggestion box programs. There are times when "yes" cannot be the answer, so as Imai wrote, "When a worker's suggestion cannot be implemented, management promptly explains why."[8]

Coaches for Coaches

Supervisors at all levels in the organization are a vital part of a successful Kaizen process, partly by coaching each Kaizeneer throughout the implementation of each Kaizen. For example, a CEO will act as a coach for their direct report, the COO, when she is implementing a Kaizen. In that way, all leaders will act as a coach of others and will also act as a Kaizeneer, seeking coaching on their own ideas from their direct supervisor. This will cascade the desired Kaizen behaviors throughout the organization. See Chapter 7 for more on Kaizen coaching.

Step 3: Implement

An idea must be implemented to qualify as a Kaizen. After identifying and discussing the idea with our supervisor and our colleagues, a person or team implements the change as part of the PDSA cycle. Keep in mind that we will test to see if the change is actually an improvement and if things changed the way we suspected in our hypothesis.

> I have been impressed with the urgency of doing. Knowing is not enough; we must apply. Being willing is not enough; we must do.
>
> **—Leonardo da Vinci**

Ideally, the person who initiated the idea should be the one to test the idea. They plan out how to implement, who to involve, and where to start. Then, they just do it (because they have already talked with teammates and their supervisor). Some Kaizens are implemented in minutes; others may take a few days or weeks or may lead to the initiation of a formal Kaizen event.

Seven Days Grace

Franciscan has learned that Kaizens should be given "seven days grace." That means everyone affected by a Kaizen gets to try it for a week before weighing in on whether it should be continued or not.

Initially, when Ronda Frieje, manager of pharmacy, launched Franciscan's Kaizen program, she experienced cases in which some people did not want to try changes that others had implemented. Ronda found that, in most cases, the Kaizens in question were effective, if allowed to work. Often the issue was that some people preferred to do things their old familiar way. Ronda created a practice she called, "seven days grace." She explained, "we have to be willing to try something new for seven days."[9] She told her staff, "I'm not afraid to go back" if the idea does not pan out. Her staff felt more comfortable going forward if they knew they could go back. Ronda had to revert only a few times in the past three years.

For example, one pharmacy Kaizen was intended to improve the productivity of assembling custom medication trays by organizing medications in locations based on the tray type. So they grouped open-heart medications together and labor and delivery medications together, and they tried it for seven days. After seven days, the analysis concluded that the productivity of tray assembly did improve, but not enough to overcome the negative effects of duplicate stocking locations that caused additional restocking effort and the greater potential of mixed medications. So they went back to the previous method of storing each medication in only one location.

Step 4: Document

Make everything as simple as possible, but not simpler.

—**Albert Einstein**

Documenting implemented Kaizens is a critical step in spreading improvement throughout the department and the organization. Figure 5.2 is a good example of a Kaizen Report. First, it was given a catchy title. Next, the nice use of pictures made the difference between the before and after conditions very obvious. It also described the before condition from the point of view of the customer—the patients, and especially in this case, their families. It nicely explained what was done to create the improvement. The effect section describes the benefit to staff and customers in a simple way. Finally, it gives credit to those to whom credit is due: the Kaizeneer and their supervisor.

Quantifying Benefits When Possible

The focus of Kaizen is not and should not be primarily on monetary gains, although monetary gains are a by-product of Kaizen. The vast majority of Kaizens improve the work or work environment and improve the customer experience, but are difficult to quantify in immediate savings or increased revenue. Again, only 6% of Kaizens at Franciscan have a quantified financial benefit that is validated by the Finance Department.

There's No Place Like Home ... Franciscan ST. FRANCIS HEALTH

Before	After
NICU can be an unexpected, scary place for our parents, so we try to decorate the beds and make it seem more warm and friendly.	I got some inexpensive foam pumpkin shapes and a paint marker to celebrate fall harvest. I can quickly write the babies' names in less than 30 seconds, and every baby in the NICU has a decoration on their bed. Parents are happy and it gives a personal touch to the bedside. I can get shapes for each season or holiday for the future.

Effect

This is a very quick and easy solution to satisfy parents and nurses, while providing compassionate care.

Name	ID #	Dept #	Supervisor	Date
Kelly Butler RN, PCC		NICU 6511	Paula Stanfill RN, MSN	10/11/08

Figure 5.2 Great example of a Kaizen report.

Kaizen would not be quick and easy if each one required a quantified dollar savings. Furthermore, it may be wasteful to always try to calculate a monetary savings for every idea. There may be a threshold dollar amount, above which it is worth the time and effort to estimate the cost savings. Franciscan works with Kaizeneers to quantify Kaizens that have the potential to save over $1,000.

> He that expects to quantify in dollars the gains that will accrue to a company year by year for a program for improvement of quality ... will suffer delusion. He should know before he starts that he will be able to quantify only a trivial part of the gain.[10]
>
> **—W. Edwards Deming, PhD**

Step 5: Share

The final step of Kaizen is to share the improvement. That means to post it, review it, and discuss it. This is done to give recognition to Kaizeneers and to spread good ideas more broadly. Different organizations use various methods for sharing, including bulletin boards, email, spreadsheets on shared drives, homegrown computer systems, or commercial web-based software.

When leaders recognize staff who initiate and implement Kaizens, there is a big increase in staff satisfaction, says Jennifer Phillips, innovation director in Virginia Mason's Kaizen Promotion Office, adding, "Staff take a lot of pride in their ideas and when they are recognized by managers at a staff huddle and thanked for their ideas and their work they feel a great sense of pride."[11]

> You are forgiven for your happiness and your successes only if you generously consent to share them.
>
> **—Albert Camus**
> *Philosopher*

The intent of this sharing, as it should be within other organizations, is to inspire your own improvements. Sometimes, sharing leads to the adoption of the same Kaizen in other units, departments, or hospitals. If people see a Kaizen they choose to adopt, that is the best approach. Better yet, if they can improve upon the idea, have them do so, and share the improvement with the original department. The rote copying of another person's Kaizen without any reflection or modification does nothing to develop one's own skills and confidence.

Over 100 examples of Kaizen improvements can be found in *Healthcare Kaizen: Engaging Front-Line Staff in Sustainable Continuous Improvements,* which contains much more detail about how to set up and manage Kaizen systems.

Sharing Kaizens: Kaizen Reports and the Kaizen Wall of Fame

Kaizen Reports are simple, single-page documents that are used by Franciscan and other health systems to document improvements, as shown in Figure 2.1. Templates like these are available for your staff and leaders to download at http://www.HCkaizen.com. Kaizen Reports are typically filled out after an improvement has been tested and implemented, but the form could also be used to first document a problem before developing an improvement. These reports are shared in numerous ways, including bulletin boards within departments, in public places, and in a web database.

A number of hospitals that Mark has worked with have implemented a process to give recognition for Kaizens, whether they come from the formal Visual Idea Board process, as introduced later in this chapter, or if they were small "just do its" that never went through the formal board process. These reports are, like the Kaizen Reports, designed to be short, concise, and to the point, capturing the essence of the improvement in a way that is understandable to a layperson. The bulletin board where these reports are displayed, Figure 5.3, was dubbed the *Kaizen Wall of Fame* by the team in the core laboratory at Children's Medical Center Dallas.

After Lisa, the new lab employee mentioned earlier, summarized her Kaizen in a report, as shown in Figure 5.4, it was posted on this wall.

Figure 5.3 Visual Idea Board in the Children's Medical Center Dallas core laboratory.

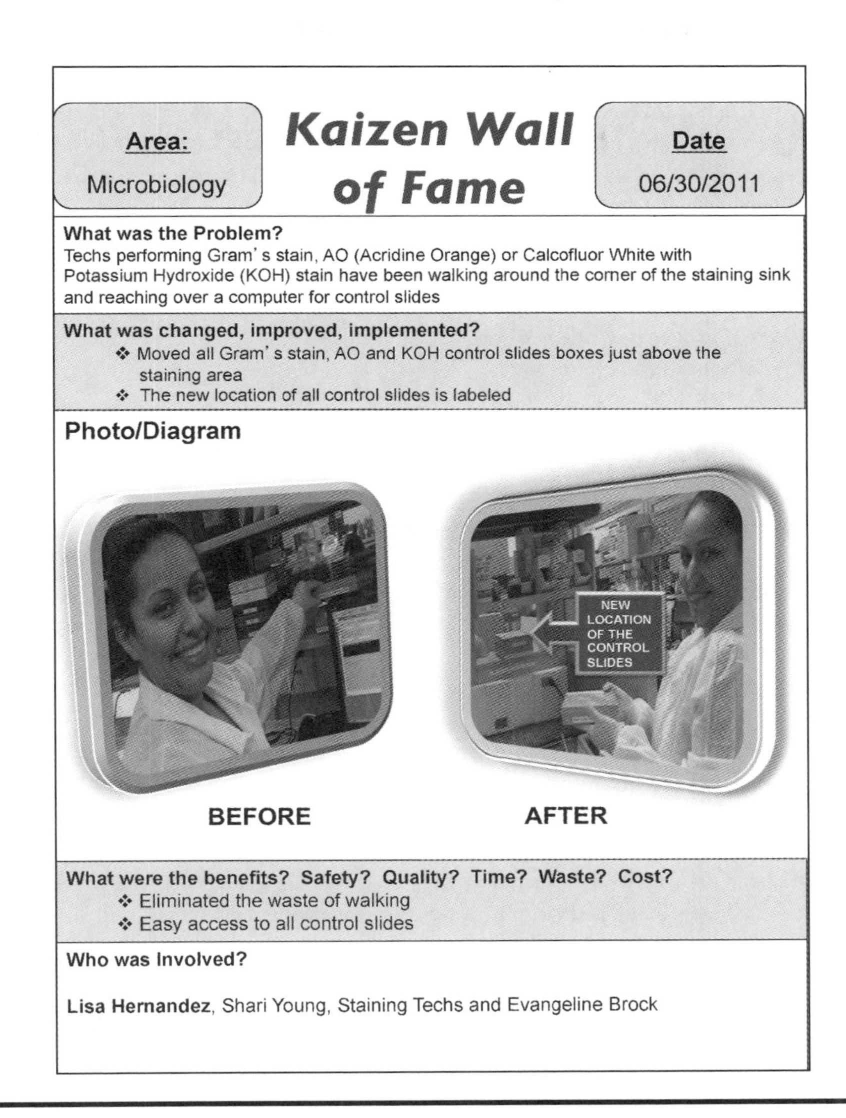

Figure 5.4 Kaizen report that was made after the Idea Card was completed.

Visual Idea Boards: Making the Entire Kaizen Visible

Another approach used by many healthcare systems is the *Visual Idea Board* approach, introduced in David Mann's *Creating a Lean Culture*, where he wrote about a "visual improvement suggestion process."[12] Visual Idea Boards can be used throughout the entire life cycle of a Kaizen—finding, discussing, implementing, documenting, and sharing improvements. In building upon the Lean concept of *visual management*, this methodology provides an alternative to the suggestion box and offers greater transparency to the improvement process. Ideas are displayed on cards on a bulletin board, allowing team members and managers to view their progress from start to finish.

In Mann's book, an idea goes through four major stages, as visualized on a bulletin board:[13]

1. Ideas are submitted.
2. Ideas are screened and advanced to a queue or rejected.
3. Ideas are actively worked on.
4. Implementation is complete.

The types of problems that have been addressed through this approach include the following (as written on cards by hospital staff):

Laboratory: "Handwriting patient information on slides takes too long and may result in identification errors."
Radiology: "The old sign on the Magnetic Resonance (MR) room barrier is hard to read. It is an old sign that looked like it was printed on an old dot matrix printer."
Telemetry Unit: "Old 3-hole punch sticks and does not punch all 3 holes consistently."
Primary Care Clinic: "Forms are stored in rolling rack on the floor; we have to bend down low to get forms."

One major advantage of Visual Idea Boards over, say, suggestion boxes, is that the current status and health of a department's Kaizen efforts are visible to all. Leaders can see the actual Idea Cards, the number of ideas, and the time elapsed since each idea was initially presented (if that date is recorded on the card).

Senior leaders, as they conduct *gemba walks* in a department, will have a visual cue about how a specific department's leadership is fulfilling their commitment to continuous improvement. For example, they might see a board that has many new ideas, but limited progress in the evaluation and implementation steps. The senior leaders can ask questions of local management, looking to determine why there seems to be limited progress.

Idea Cards

Using a structured form or card can help create standardized work for the Visual Idea Board process and can serve as a guide for PDSA thinking and problem solving, leading to more effective improvement.

Either way, the important thing is that a frontline staff member raised the issue to be discussed. In an old suggestion box-type program, managers might simply say, "no, we don't have the budget," and that would be the end of the discussion. In the

Kaizen approach, everybody works together to find a way of addressing the problem that was brought up. We do not have to always implement exactly what was submitted, but it is the role of a manager to help find a countermeasure that works.

The presence of a board (or software) does not guarantee that ongoing Kaizen activity will occur. One hospital had an initial wave of improvements that were managed through boards in multiple nursing units. Over time, enthusiasm seemed to wane, and the Visual Idea Boards were eventually reclaimed as general bulletin board space. A nurse from the original team that created the board looked back and said, "The managers didn't go around asking for ideas," because they had too many other demands on their time. In other hospitals, the success of the Kaizen program has been a factor of the amount of effort put in by the supervisors and managers in each department. One cannot just put up a board and expect to get results from Kaizen if one is not willing to put in daily effort, as managers, to encourage ideas and their implementation.

Visual Idea Boards provide an open and transparent alternative to the traditional suggestion box. Structured cards prompt staff members to follow the PDSA improvement process for defining a problem, suggesting a countermeasure, and then testing that change in a scientific way. The boards allow staff and leaders to hold each other accountable for the progress of cards, leading to faster improvement cycles and improved team communication about ideas that can improve patient care and make people's work easier.

Electronic Kaizen Systems: Making Kaizen More Broadly Visible

> Our online Kaizen database so improved my ability to do my job well that I would highly recommend starting a Kaizen program with an online database on day one.
>
> **—Julia Dearing**
> *Franciscan regional Kaizen coordinator*

In recent years, many organizations have developed electronic online databases to help facilitate their Kaizen program. The evolution of many organizations is similar. For example, when Franciscan started their program, all Kaizens were handwritten. Shortly after the start of the program, they introduced a Microsoft

PowerPoint template that could be filled out and emailed to the supervisor for approval. Franciscan also had its information services department create a special email box for people to submit Kaizens electronically. Over time, email became the main way Kaizens were turned in. Then, in 2009, a homegrown web-based database was introduced. By the start of 2012, nearly 95% of Kaizens were recorded using the online database, 3% by email, and 2% by paper.

Advantages of an Electronic Online Database

The electronic online database enhanced reporting, tracking, and sharing of ideas. Other key advantages of the online database over paper forms at Franciscan have included the quick entry and automatic routing of ideas, the ability to hold ideas for later, giving electronic approval, and the quick search and retrieval of previous ideas.

Quick Entry and Categorization

The Franciscan system allows for the quick entry of ideas, capturing basic information, such as the title, before condition, after condition, and the effect. Ideas are also categorized upon entry in the database when there is not a clearly quantifiable benefit. Finally, the Kaizeneer and all contributing Kaizeneers' names are added, and any attachments are submitted, such as supporting pictures or other files.

Automatic Routing and Electronic Communication

Another advantage of an online database is that completed Kaizen Reports can automatically be routed to the proper supervisor, where she can either approve the Kaizen or convert the Kaizen to an idea to be implemented later.

Once the supervisor views it and any attached files and verifies the Kaizen is complete and meets the criteria for approval, she can print the Kaizen to be posted, and then the Kaizen can be approved. The Kaizen is automatically flagged for the regional Kaizen coordinator to review. Then, once approved, the Kaizen is immediately available for anyone in the organization to search and view. An email notice is sent automatically to the employees who submitted the Kaizen after the online approval process is completed.

Quick Search and Retrieval

Kaizen Reports can be quickly searched and retrieved. Kaizeneers can quickly search the status of all Kaizens that they have submitted electronically. The search can be done by department, submitter's name, title, or description. Users can view or print any of the Kaizens found.

Electronic Kaizen within Intermountain Healthcare

At Intermountain Healthcare's Utah North Region, an electronic *Idea Tracker system* was put in place, as their tactical applications team made modifications to an existing online event reporting system. They use a Visual Idea Board approach, and a unit clerk then enters completed ideas into their electronic tracker.[14] The goal is to make it easier for people across their three sites to search for ideas and share information. Another goal was to be able to roll up statistics about their Kaizen efforts, financial and otherwise, across departments for a more global view of their program and its health.

Electronic Kaizen at Vanderbilt

In 2005, Gregory Jacobson, MD, an emergency medicine physician and faculty member at Vanderbilt University Medical Center (VUMC), was given a copy of Imai's *KAIZEN* by his department chair, Corey Slovis, MD. After reading Imai's book, he decided that the emergency department (ED) was a perfect setting to engage the residents and teach them Kaizen principles as a formal method for quality improvement.[15]

Later that year, after educating the residents and faculty physicians on the history and principles of Kaizen, Jacobson and a coworker, Richard Lescallette, built a web system that they called the *Kaizen Portal*, which later evolved into the improved Kaizen Tracker. This was the place where any emergency medicine physician or employee could submit an improvement idea.[16] All comments about a Kaizen were kept in the Tracker, which could be viewed in the web application by anybody, instead of getting lost in people's email inboxes.

Unlike the organizations mentioned previously, the Vanderbilt ED started their program with an electronic idea submission tool. As Jacobson reflected, "I just couldn't have imagined doing this with a paper system because, in the hospital and ED, we were solving many other problems with technology, including the adoption of electronic order entry and the digitization of the radiology department," adding that "you couldn't walk five feet in the department without bumping into a computer." Additionally, there were 300 staff members working in a 75-bed department that was spread across two physical locations, and "if we would have had a bulletin board, I wouldn't have known where to put it."[17]

As a Kaizen was being discussed and implemented, the author could see and participate in discussion about that idea. When improvements were finalized, a notification was sent to the entire department, to keep people informed about changes. The department was also notified if an idea could not be implemented, creating an important feedback loop to "assure that no concerns are simply dismissed without explanation or discussion."[18] Completed Kaizens were also categorized and kept in the Tracker to be viewed or searched by all ED staff.

The Tracker was used continuously in the ED until 2011, when they converted their system to a newly available commercial version of the web-based application from a startup company called KaiNexus™ that was cofounded by Jacobson after licensing the technology from Vanderbilt. The new software increased the visibility and collaboration beyond a single department, as they expanded its use beyond the emergency department, while adding improved metrics and reporting.

KaiNexus offers an enterprise-level improvement software platform designed to help support a Lean and Kaizen program for an entire medical center by supporting staff and supervisors in collecting, routing, tracking, approving, archiving, and communicating thousands of Kaizens. Jacobson says, "KaiNexus isn't intended to replace team huddles and face-to-face interaction," adding, "using email to manage Kaizen usually fails, but a web-based system is more effective because it's inclusive, transparent, collaborative, and nothing gets lost."[19] Coauthor Mark is the chief improvement officer of KaiNexus.

Jacobson and his coauthors analyzed their Kaizen results for the adult ED in a journal article published in 2009. One notable achievement was their success in engaging residents in Kaizen, as 93% of residents submitted at least one improvement idea. Overall, there were 408 opportunities for improvement recorded in the adult ED through the Portal and the Tracker systems over a span of 56 months. Of these ideas, 43% came from residents, 40% from faculty, and 17% from other staff.

> We no longer have a "complainer" type of culture anymore; if there is a problem, our department's culture now maintains a "don't complain, send a Kaizen" attitude. Our residents and faculty know that feedback occurs in hours to a day or so, and the problem will be fixed, addressed, or an explanation provided on the issue they have raised.
>
> It has changed my life. This department is dramatically improved because of it, and our residents and faculty are much more content. It has also increased efficiency as problems are fixed quicker, and finally it allows leaders and managers to be held accountable.
>
> —**Corey Slovis, MD**
> *Chairman of the Vanderbilt Department of Emergency Medicine*

As of April 2013, the KaiNexus™ software is being used in multiple departments within Vanderbilt University Medical Center and is being implemented at over fifteen other major medical centers.

Conclusion

Regardless of the exact mechanisms, paper or electronic, used through the process, the five basic steps to Kaizen are:

1. **Find** an opportunity to change something for the better.
2. **Discuss** the opportunity and potential improvements with your team members and your supervisor.
3. **Implement** the best improvement.
4. **Document** a Kaizen Report.
5. **Share** the Kaizen Report with others.

The real power of Kaizen occurs over time when everyone in an organization is empowered to apply these five steps to improve their daily work, developing their skills and abilities to make improvement. Nancy Mosier, former manager of pediatrics at Franciscan, said, "Kaizen has made our staff look at their work environment more closely, to see their workflow from a new perspective, and it has empowered them to make their own changes." To Nancy, the specific process of Kaizen is not as important as what Kaizen has enabled her staff to do. One afternoon, Joe walked with Nancy around her unit as she pointed out the hundreds of Kaizens she and her staff have done over the last few years. Joe agreed with Nancy when she said, "They might seem like little things, but they have added up to make all the difference."[20]

Discussion Questions

- Is it always a good strategy to start with small Kaizens first that make your work easier?
- What are some of the advantages and disadvantages of having an electronic system for managing and sharing Kaizens?
- Why is it important for people to document and share completed Kaizens?
- Do your supervisors have the disposition to discover how to say "yes" to Kaizens?
- Do your supervisors have the skills to coach Kaizeneers? If not, how can the organization support and develop them?

Endnotes

1 Eckhardt, Joy, personal interview, July 2011.
2 Garner, Bernice, personal interview, July 2011.
3 Bodek, Norman, and Bunji Tozawa, *How to Do Kaizen: A New Path to Innovation* (Vancouver, WA: PCS Press, 2009), 6.
4 McRoberts, Susan, personal interview, July 2011.

5 Mosier, Nancy, personal interview, July 2011.
6 Pettigrew, Laura, personal interview, August 2011.
7 Imai, Masaaki, *KAIZEN: The Key to Japan's Competitive Success* (New York: McGraw-Hill, 1986), 114.
8 Imai, 115.
9 Frieje, Ronda, personal interview, June 2011.
10 Deming, W. Edwards, *Out of the Crisis* (Cambridge MA: MIT CAES Press, 1982), 123.
11 Virginia Mason Medical Center, "Using Lean Ideas in Our Everyday Work," http://virginiamasonblog.org/2013/01/30/using-lean-ideas-in-our-everyday-work/ (accessed January 31, 2013).
12 Mann, David, *Creating a Lean Culture* (New York: Productivity Press, 2005), 146.
13 Mann, *Creating a Lean Culture*, 147.
14 Sellers, Bart, personal interview, September 2011.
15 Jacobson, Gregory, personal interview, September 2011.
16 Jacobson, Gregory H., Nicole Streiff McCoin, Richard Lescallette, Stephan Russ, and Corey M. Slovis, "Kaizen: A Method of Process Improvement in the Emergency Department," *Academic Emergency Medicine* 16, no. 12 (2009): 1342.
17 Jacobson, personal interview.
18 Jacobson et al., "Kaizen: A Method of Process Improvement," 1343.
19 Jacobson, personal interview.
20 Moiser, Nancy, personal interview, July 2011.

Chapter 6

The Role of Senior Leaders in Kaizen

Quick Take

- CEOs and senior leaders need to participate in the establishment of a Kaizen program and cannot delegate the creation of a culture of continuous improvement.
- Some key behaviors are required of leaders at all levels and need to be modeled by senior leaders, including believing and participating in Kaizen, asking others to participate, developing people, recognizing and rewarding staff, sharing Kaizens, and selling the benefits of the program.
- Senior leaders need to communicate expectations, ensure adequate resources are available, sponsor the recognition program, share Kaizens, and thank people personally.

> It is not enough that top management commit themselves for life to quality and productivity. They must know what it is that they are committed to—that is, what they must do. These obligations cannot be delegated. Support is not enough; action is required.[1]
>
> **—W. Edwards Deming, PhD**

The Reluctant CEO

John Toussaint, MD, was CEO of ThedaCare from 2000 to 2008. In 2002, he initiated their Lean efforts under the heading of the ThedaCare Improvement System (TIS).[2] In their early years, much of the focus was placed on weeklong Rapid Improvement Events, or RIEs.

As CEO, Toussaint promoted Lean, while sponsoring and encouraging improvement through events. He "explicitly communicated to senior leaders that they were expected to participate in at least two RIEs each year, making it clear to all other leaders and managers that this meant them too."[3] Looking back, Toussaint reflected, "It took commitment and focus at the senior level to clarify that TIS was the beginning of a cultural transformation, and that everyone would be affected."[4]

Toussaint, however, was not yet directly participating in events. One of his mentors was George Koenigsaecker, a member of *IndustryWeek*'s manufacturing hall of fame for his Lean success as a manufacturing company president and investor. Koenigsaecker asked him if he had been part of an RIE team and Toussaint answered, "No, but I am trying to schedule one." After Koenigsaecker repeated the same question for a few months, Toussaint finally participated in an event.

Koenigsaecker then asked, "Did you stay for the entire time on Monday, or did you have to go to some important meeting halfway through and miss the mapping of the initial state?" Toussaint admitted that he had left the event on Monday for two hours and that he had left many times throughout the rest of the week, excusing his absences as "the fire fighting that most CEOs have to do."[5]

The response from Koenigsaecker was blunt, telling Toussaint, "You're never going to learn this unless you tell your secretary you are on vacation for this week, which means you are at the entire event, participating with the staff, learning the tools, and understanding what it takes to improve a process."

Toussaint then fully participated in the next event, better understanding the ups and downs of the group's emotions during the week, giving him a better appreciation for what is really involved in Kaizen. Toussaint reflected that "the staff and physicians became very curious as to why I would take a week off of my duties and work on the obstetrics unit or on the loading dock at home care. They realized we were serious about this improvement thing the more they saw me working side by side with them to fix problems or on my weekly gemba visits to their units."

Over time, Toussaint participated in 14 different weeklong events, including value stream assessments, and Lean design exercises called *3Ps* and *2Ps*. Why was it important for the CEO to participate? Toussaint reflected on this, saying, "When I went to the floor as a team member on an improvement activity, it sent a loud message first that I cared about the plight of the frontline workers and that this wasn't some fad or management trick."[6] He adds, "The journey required that I behave differently."[7]

Leadership also plays an important role in ThedaCare's *continuous daily improvement* activities, which are expected to account for 80% of their overall improvement. Toussaint and other senior leaders spend time in the gemba daily, as does current CEO Dean Gruner, MD, asking questions and encouraging all forms of improvement. Rachelle Schultz, CEO of Winona Health (Minnesota), requires all leaders to participate in a minimum of three Kaizen events each year "so that they have firsthand experience with the work and the behaviors that needed changing."[8]

Key Actions for Leaders at All Levels

> The key to successful Lean implementation is that leaders have to change. We have to change from the all-knowing, being "in charge," autocratic "buck stops with me," impatient, blaming person—who is a control freak—to the person who is patient, knowledgeable, a good facilitator, willing to teach, actually willing to learn, be a helper, an effective communicator, and be humble.[9]
>
> **—John Toussaint, MD**
> *CEO, ThedaCare Center for Healthcare Value*

In an effective Kaizen program, leaders at all levels must be actively engaged. First-level managers are involved in every Kaizen, coaching and guiding staff. Middle-level managers drive Kaizen by showing their commitment, determination, and giving guidance, so everybody is working on Kaizen. Senior leaders see to it that Kaizen becomes a part of the organization's culture by first modeling and teaching these behaviors. Leading this change "requires new habits, new skills, and often a new attitude throughout the organization, from senior management to frontline service providers."[10] Rachelle Schultz, CEO of Winona Health (Minnesota), says the role of senior leaders is to "model, support, and coach the new behaviors" that are required for Kaizen, "putting intentional changes in leadership work, including required time in the gemba, because it is too easy to fall back into old habits."[11]

At Franciscan, everyone is continually reminded how much influence leaders have. The departments that are performing the best in Kaizen are those with leaders who believe in Kaizen and decide it is important enough to incorporate into everything they do. They made a strong commitment to the Kaizen process, and it shows in their results. Senior leaders need to model the behaviors they want to see throughout the organization, as discussed below.

Key Action 1: Believe in the Power of Kaizen

Revenue management is a top-performing department at Franciscan. According to their director, Lynne Meredith, it starts with believing in what Kaizen can do for the department. She says, "If you don't believe in Kaizen, your managers won't believe, and your staff won't believe." When Lynne noticed someone make a change, she would say, "Hey, that's a Kaizen. Make sure you write it up." Lynne also noticed that her employees needed a regular reminder of why Kaizen is important and what's in it for them. She says, "Part of believing in Kaizen is having a ready answer to the question about why we're doing Kaizen, and it cannot be because administration wants us to do Kaizen. Rather, the reason we do Kaizen—all of us—is to recreate our world around us and continually mold and transform it into the way it could and should be, and that is powerful."[12]

Toussaint writes, "Unleashing the energy and the ideas of every employee to identify and solve problems is a huge task, but it changes everything. Frontline staff are typically excited when their ideas for improvement are implemented. They brag a little to friends, and the energy goes viral."[13]

> Saying yes to kaizen was easy for me. As a leader, I recognize my role is to lead others to create the future of choice for our organization. I know our future success depends on the continuous development of the process improvement skills and abilities by all our staff in order to enhance the services we provide to our patients and their families.[14]
>
> **—Bob Brody**
> *CEO, Franciscan St. Francis Health*

Key Action 2: Participate in Kaizen

Imai's expression "Kaizen is for everybody" includes leaders at all levels. This does not mean jumping in and solving other people's problems. It means modeling Kaizen behaviors by initiating and conducting your own Kaizens, making small improvements in your own work before you ask others to do Kaizen. Doing your own Kaizens shows employees that you are serious about improvement, and you can use your experiences to coach them and relate to them. If it was difficult to do your first Kaizen, be honest and share those experiences with your team—they will appreciate it.

Keith Jewell, Franciscan COO, says, "Major change requires a senior leader champion and their significant involvement." He continues, "It surprises me how much people look to the leader. Staff want to know that the leader believes in the Kaizen program. So, I role model what I expect and what I want others to model. It accelerates the success of the program. If they see my Kaizens in the database, they know that it is important."[15] Figure 6.1 is a Kaizen that Keith shared with the organization, showing that improvement started with him and that he could have fun in the process.

> The [servant] leader has a vested interest in the success of those being led. Indeed, one of our roles as a leader is to assist them in becoming successful.
>
> **—James C. Hunter**
> *Author*

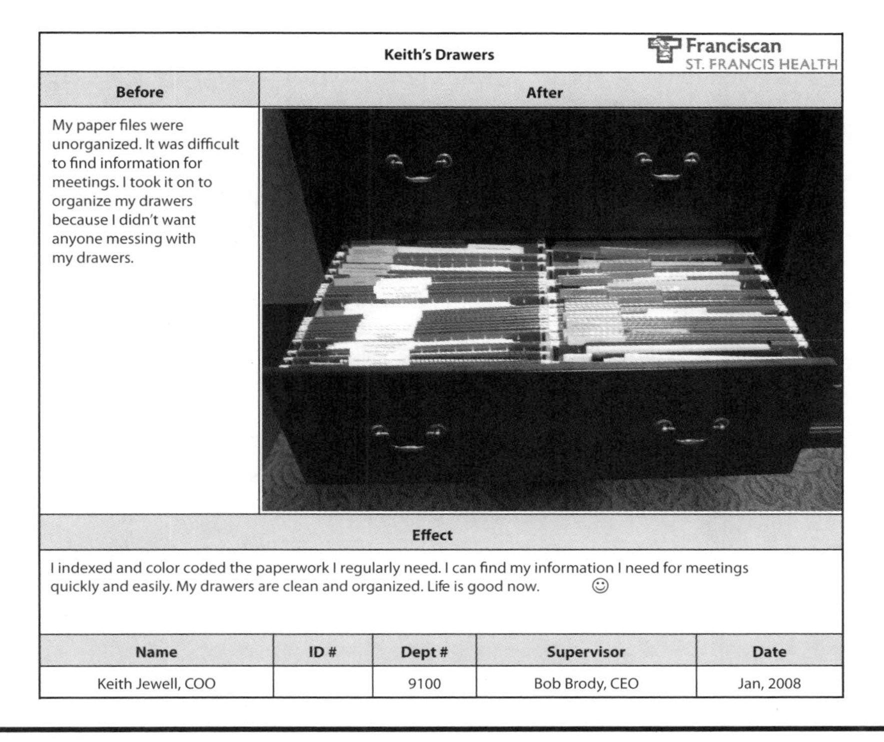

Keith's Drawers — Franciscan ST. FRANCIS HEALTH

Before	After
My paper files were unorganized. It was difficult to find information for meetings. I took it on to organize my drawers because I didn't want anyone messing with my drawers.	

Effect

I indexed and color coded the paperwork I regularly need. I can find my information I need for meetings quickly and easily. My drawers are clean and organized. Life is good now. ☺

Name	ID #	Dept #	Supervisor	Date
Keith Jewell, COO		9100	Bob Brody, CEO	Jan, 2008

Figure 6.1 A Kaizen report created by the chief operating officer.

Another important aspect of Kaizen participation is to be a *servant leader*. Again, you want to empower your employees to make their own improvements, but you need to take an active role in evaluating and giving feedback on ideas, as well as following up on their implementation. "The manager's role is not to solve problems, but to coach and mentor staff in problem solving," says Toussaint.[16]

There will be many cases when a staff member or lower-level manager raises an issue that requires help from others. In those cases, leaders need to escalate the opportunity to the right level of leadership or the right department. In a traditional organizational model, the leaders who are more senior get to make more decisions. In a Kaizen approach, this is not the case, because leaders push decision making down in the organization as much as possible. The more senior leaders tend to have a broader view of the organization or a wider managerial span, so they can help fix problems that frontline staff cannot solve on their own.

Key Action 3: Just Ask

Starting a Kaizen program can be as simple as asking staff to do Kaizen. Keith Jewell, Franciscan COO, said, "Kaizen has something in it for everybody. I get an economic return, in the form of savings, which far exceed the investment in the program. The staff gets to make their work easier and better. We get improved service and quality for our patients. It is a program that benefits all of our customers. It really is a win–win program. There is not a lot of controversy in doing good things for others. So, I just ask staff to participate. We don't require it, but we do encourage them to participate."[17]

Evangeline Brock, a medical technologist at Children's Medical Center Dallas, says, "Five years ago, our ideas weren't listened to." Brock said, "New employees would ask why we do things a certain way, and there wouldn't be a good answer."[18] As the lab's leaders embraced Kaizen, they found it was necessary to continually reinforce to staff that "it's OK to speak up," because people are often very conditioned against doing so. Asking once is probably not enough; managers need to ask every day.

> As leaders, we are preoccupied with putting the message out and not preoccupied enough with listening. Listening is a way we show "respect for people." What we hear and how we react is a critical piece of the PDSA cycle.[19]
>
> **—Michel Tétreault, MD**
> *CEO, St. Boniface General Hospital*

Leaders have to be willing to make time available to participate directly in Kaizen. Jennifer Rudolph, formerly of Park Nicollet Health Services, emphasizes to their leaders, "If you approve the idea, then you are approving the time to work on it."[20]

Another large hospital's CEO had initiated a program where staff members were encouraged to call out problems to senior leaders, following the mantra of "no problems is a problem." Unfortunately, so many problems were called out, it overwhelmed the ability of leaders to react and help. The CEO later offered his apology for raising expectations for improvement and then disappointing people. The story illustrates why it is important for leaders to ask people to not only call out problems, but to also ask them to think of improvements they can implement locally. The story also illustrates why an organization might want to pilot a Kaizen program in a single department rather than starting as an entire hospital, to learn the amount of time and support that is really required.

In a well-known Toyota story, a Japanese leader told a North American manager that "no problems is a problem," emphasizing that a culture of openness and transparency is a critical starting point for improvement.[21] Hiding problems interferes with improvement and the prevention of future errors. An Agency for Healthcare Research and Quality (AHRQ) study suggested more than 50% of healthcare professionals did not report any medical errors over a 12-month period due to a blame-and-shame culture.[22] Furthermore, in a survey of 1,000 incident reports, 89% were submitted by nurses, with 2% coming from doctors;[23] 60% of these incidents were deemed preventable.

Ask, Don't Tell

Another general tenet of Kaizen is to "ask, don't tell." In general, people do not like being told what they should and shouldn't do. In Franciscan's Kaizen program, it is optional for staff to participate. Keith Jewell noted, "It wasn't forced, and therefore we didn't experience a lot of resistance. Everyone was charged up by the potential opportunity."[24]

Whether you are doing a Kaizen yourself or coaching someone else through one, a good way to avoid telling people what to do is to share your thoughts and ideas in question form. When Joe Click, pharmacy supervisor at Franciscan, works with staff on Kaizen, he says, "I avoid telling people what to do. Instead, I do my best to ask questions, such as, 'what do you think of this idea? How can we implement it?'"[25]

As David Mann writes in *Creating a Lean Culture:*

> The classic sensei (teacher) is Socratic in approach, teaching by stretching the student's thinking and perceptions through questions that stimulate the student to consider entirely new possibilities.[26]

John Toussaint, MD, former CEO of ThedaCare, shifted from being the leader who had all the answers to one who asked a lot of questions. To help staff see waste and drive improvement, Toussaint was successful by "asking staff questions, which helps staff to think about the problems themselves. The leader should never tell the staff what to do, but they instead need to be able to ask the right questions."[27]

Questions should be asked humbly, with respect, and in a thoughtful and careful way. For example, you could ask, "Could you help me understand how this process works?" Kaizen leaders should avoid asking questions they know the answer to, as if to push or manipulate the person to give a certain answer. These should be questions of inquiry or questions that get the other person to think in a new way.

Senior leaders should be aware that asking questions may cause anxiety or lead to staff feeling like they are being criticized. If senior leaders have not been present and visible in the workplace, it may take time to build trust amongst managers and staff in various departments before effective dialogue can occur.

> Be careful of directives that are cleverly disguised as questions. When Mark's wife asks him, "Why did you throw your socks in the middle of the floor?" they both laugh about how she is not really

asking a question of inquiry. She is not searching for a root cause of Mark's behavior. Her question is really a statement that says, "Please don't leave your socks on the floor."

Key Action 4: Use Kaizen to Develop People

Part of empowerment is allowing your staff to make mistakes, as discussed in Chapter 2, and using those mistakes as opportunities for learning. That said, we do not want to allow reckless mistakes or changes that would obviously hurt people. However, it is also the leader's job to ensure the learning environment is emotionally safe for staff, being cautious with people's physical safety without being overcautious about every small change or needing to have zero risk of failure. There are rare instances when somebody might truly have a bad idea that would cause harm. Instead of just rejecting the idea, take the time to develop the improvement skills of your staff by allowing them to think the risks through, under your guidance, and empowering them to find a better solution.

Key Action 5: Ensure Staff Members Are Recognized and Rewarded

> The recognition through the Kaizen program has made me feel valued and lets me know I am making a real difference.
>
> **—Craig Whitaker**
> *Financial analyst, Franciscan*

At Franciscan, Lynne Meredith wants her staff to recognize that it is important to her that they take time out of the day to focus on Kaizen. So, she regularly recognizes people in her staff meetings and makes a point of explaining how those people went out of their way to coordinate with another in order to implement a particular Kaizen. Lynne said, "I have been persistent in reinforcing the Kaizen behaviors that I have wanted to see more of."[28]

Recording an improvement and displaying the results allows the person or team to receive the recognition they deserve. At Franciscan, managers and supervisors post Kaizen Reports locally in their departments, and selected ones are posted prominently in high-visibility areas, such as in hallways or the cafeteria. Ronda Freije, pharmacy manager, shares graphs with her staff showing how her department is performing in the organization. She said her staff "are proud that they are one of the top-performing departments in total Kaizens turned in. It is a way for our techs to get recognized. It gives them a sense of achievement." Ronda also has potluck lunches where she recognizes recently turned-in Kaizens, and she occasionally brings in ice cream as a way of showing her appreciation to the team.

Children's Medical Center Dallas and the other hospitals that implemented this method with Mark typically did not give financial rewards or direct incentives to participate in Kaizen. The recognition and reward for Kaizen, beyond making one's work easier or the good feeling that came from helping colleagues or improving patient care, came from the wall itself. The names of people involved in the Kaizens were all displayed on the wall. This gave people recognition amongst their peers, their managers, and senior leaders who visited the department to review the improvements that were made.

Key Action 6: Share and Spread Ideas

The real power of Kaizen occurs when ideas are shared. Sharing enables people to benefit from others' ideas, allowing them to realize that they too can make similar improvements to their work.

At Franciscan, success comes by keeping the practice of Kaizen in front of staff. The most successful Kaizen leaders devote a portion of every staff meeting to recognizing individuals, asking about recent Kaizens, and sharing ideas. They utilize staff meetings to discuss Kaizens that will affect their entire team. They also share regularly in other venues, such as rounding huddles. At Franciscan, one Kaizen saved a few thousand dollars by evaluating what items nurses placed in the sterile field during surgeries and deciding to hold back some less frequently used items until the surgeon asks for them. After the Kaizen was shared widely, Franciscan received dozens of Kaizens over the next few months based on the same idea, with one saving $57,000 per year.

At Children's Medical Center Dallas, ideas from the core laboratory have been seen by other staff members on their Kaizen Wall of Fame, as shown in Chapter 5. In one instance, some laboratory information system comment templates were adjusted so they could be opened with fewer keystrokes. This time-saving improvement was shared and adopted throughout the other parts of the lab, including microbiology and virology.

Key Action 7: Sell the Benefits

Leaders at Franciscan regularly spend time reminding staff of the benefits of Kaizen. They communicate those benefits using examples from throughout the organization. They communicate that Kaizen enables compassionate concern for patients and their families by improving safety, quality, and customer satisfaction. Leaders also reinforce that Kaizen enables organizational stewardship by improving productivity, lowering costs, and increasing throughput. Finally, leaders remind their employees that Kaizen demonstrates respect for their fellow employees by empowering them to make a difference. Collectively, these benefits enable their organization to be financially stronger, which leads to increased job security for all staff.

Tétreault, CEO of St. Boniface, connects Kaizen to the four true north objectives of the organization: "satisfy patients, engage staff, do no harm, and manage resources—in that order." In talking with nurses, he emphasizes the need to reduce the occurrences of patient harm and "the nurses nod vigorously." It has "become [St. Boniface's] belief" that if you focus on the first three objectives, without ignoring cost altogether, that the fourth objective "basically takes care of itself."[29]

The Specific Role of Senior Leaders

Senior, or top-level, managers see to it that Kaizen becomes a part of the culture of the organization. Franciscan considers the top-level managers to include the C-level, the president, the vice presidents, medical directors, and executive directors.

A healthcare organization's board can play an important role in driving culture change and holding senior leaders accountable. "Having a knowledgeable and supportive board is critical," especially if senior leaders "lose their courage because the going gets tough," writes Toussaint.[30] Tétreault says his board, at St. Boniface General Hospital, has told him, "What's important [with Lean and Kaizen] is that the culture is changing."[31]

Leadership and Kaizen Participation Starts at the Top

James Dague, retired CEO at IU Health Goshen Hospital, says a critical factor in the success of their Kaizen system was his high visibility, because he made rounds one to three times a week, and "the constancy of it has to be there at the top level." Dague asked about improvements that people are working on and how their work perpetuates the hospital's mission and their personal missions. Dague humbly emphasizes that, while he played an important role, his employees deserve the credit for their program, as he says "they didn't have to do [so many improvements]."[32]

Going to the Gemba

At Iowa Health System, Ray Seidelman, manager of performance improvement, says that for daily Kaizen to be successful, "it needs to be the front-line staff identifying the problems and initiating improvements."[33] Yet, senior leadership has an important role to play in driving people to improve by asking challenging and provocative questions to staff members, says Gail Nielsen, their director of learning and education. For example, a senior leader saw that HCAHPS (Hospital Consumer Assessment of Healthcare Providers and Systems) scores were down, so she challenged staff to find ways to improve. Rather than blaming people, senior leaders at Iowa Health System went to the floors (the gemba) to spend time with staff and patients, to see what the real situation was, and to get ideas from staff members. Nielsen says it comes back to top leaders asking, "I wonder what the real situation is. I wonder what's going on?" This style of leadership is very different than a traditional command-and-control model, where a senior leader might mandate improvement without getting directly involved other than continuing to look at the scores.

Gary Kaplan, MD, CEO of Virginia Mason Medical Center, "tours the hospital daily looking for problems and solutions," while "[e]veryone is encouraged to look for changes to make work more efficient."[34] ThedaCare's Gruner adds that going to the gemba connects him to the work the health system does, and helps him model the behaviors that he expects from his staff, saying, "Colleagues enjoy the time I spend with them as they know how precious time is and that spending it with them signals the importance of their work."[35]

Key Actions for Senior Leaders

In addition to the above behaviors, there are specific roles that senior leaders must play to ensure the success of a Kaizen program, as discussed below.

Key Action 1: Communicate Expectations, Prioritize, and Set Direction

The CEO needs to communicate the expectation that the organization wants all employees and affiliated staff to participate in Kaizen. Ensure that it is communicated several times each year in multiple media formats. Communicate the long-term expectation; then ask people to start slowly, such as expecting one or two improvements per person per year. Then, increase the expectation each year as appropriate, based on the amount of improvement the overall system can support.

> The Kaizen philosophy assumes that our way of life—be it our working life, our social life, or our home life—deserves to be constantly improved.[36]
>
> **—Masaaki Imai**

At Franciscan in 2007, CEO Brody and COO Jewell challenged each employee to submit two Kaizens per month as a long-term goal. In order to get employees used to the practice of Kaizen, they asked all full-time employees in 2008 to submit at least one Kaizen that year. Then, in 2009 they asked full-time employees to submit at least two. Employees were not punished in any way for not reaching that goal.

At the Utah North Region of Intermountain Healthcare, Bart Sellers, regional manager of management engineering, says that senior leaders set no expectations on the number of Kaizens for the first few years of their program. The initial goal was to figure out how to do Kaizen, including the roles and expectations for managers. Eventually, a modest goal was set at one Kaizen per employee per year. Sellers says, "Before setting the expectation, we had some departments taking off and others were ignoring it," adding that, "setting the measure was good for awareness and getting the attention of managers," but the goal would not be used in a punitive way.[37]

Beyond setting goals for the number of Kaizens, senior leaders are also in a position to help prioritize improvements, by selecting events or system Kaizen projects that best serve the strategic needs of the organization. For example, at the Cancer Treatment Centers of America, the C-level leaders select high-level projects, such as RIEs and Six Sigma projects, while also clearing paths and providing strategic support for the smaller A3 improvements that are primarily generated by staff members.[38] "The beauty of the A3 program is that it empowers front-line stakeholders to select the projects that will make the biggest impact for their patients," says Jennifer Smith, their director of Lean and Six Sigma.[39]

Generally, staff members and first-level managers will choose their daily Kaizen activities based on local pressures and challenges, but the communication from senior leaders about vision, priorities, and challenges will help people point their Kaizens in the right direction to meet broader goals. At Franciscan, leaders communicate the organization's annual goals through a program called *Journey to Excellence*. Leaders and staff are encouraged to focus their Kaizen activities around the 7 to 10 goals for a given year.

Key Action 2: Ensure Adequate Resources Are Available

> Kaizen requires an upfront investment of time and effort, but it pays you back at least ten times the upfront investment in patient and staff satisfaction and retention.[40]
>
> **—Paula Stanfill**
> *Manager, NICU, Franciscan St. Francis Health*

An under-resourced Kaizen program will underperform. In addition to top leadership support, an organization needs to devote some budget and some people to help facilitate the program. In Chapter 8, we share the staffing levels and organizational model that Franciscan has found to work well for their Kaizen Promotion Office.

Senior leaders are also in a position to ensure that other leaders have enough time for Kaizen on a daily basis. At Toyota, a person in a *team leader* role has between 5 and 7 direct reports,[41] about the same ratio of nurses with a charge nurse in a medical or surgical unit. Like a charge nurse, the team leader is a worker who provides coaching and support, but can jump in to do the actual work, if need be. In the first level of Toyota's salaried management, a *group leader* might have 4 team leaders reporting to them.[42]

In comparison, many hospitals (as well as organizations in other industries) may have eliminated a layer of management to reduce their costs. A common side effect is that managers end up having such a broad span of responsibility (such as 25 or 30 employees on a shift) that they cannot devote enough time to coaching everyone. A few years into its Lean journey, one leading hospital added a layer of management back in, learning it was required to properly support improvement efforts. Increasing costs in that category allows them to all work more effectively in reducing overall system costs through various types of Kaizen.

Key Action 3: Sponsor a Recognition and Incentives Program

It is important to ensure that employees are recognized for completed Kaizens. For many people, getting public credit and recognition is more important than financial rewards. Simply putting Kaizen Reports on a bulletin board and thanking people in team meetings can create a lot of goodwill and pride in people's improvements. Senior leaders also need to provide the budget and support required for a financial reward system for employees, if one is desired. See Chapter 7 for a discussion of the pros and cons of using Kaizen participation as part of an annual performance review process.

Key Action 4: Share Notable Kaizens

At Franciscan, the COO forwards a *Notable Kaizen* to all 4,000 employees once or twice a month. The Kaizen is shared with a sentence or two from the COO, which is typically prepared by the Kaizen program staff for his final review and decision about sharing.

Here is one of the Notable Kaizens that have been sent at Franciscan:

> Dear Franciscan Family,
> This Kaizen demonstrates our value of Christian Stewardship. Two of our Radiology Technicians took the initiative to reduce the cost of handouts for our patients. They involved others in their department along with a physician to ensure the new handout was correct and complete. Now the same basic information is being communicated in a less expensive way (Figure 6.2).
>
> **—Keith Jewell, Sr. VP and COO**

Key Action 5: Thank People Personally

In the course of rounding or going to the gemba, it means a lot to frontline staff to have a senior leader stop and acknowledge their Kaizen work. A sincere question about improvements, a smile, and a handshake will go a long way toward giving recognition and encouraging people to do more Kaizen.

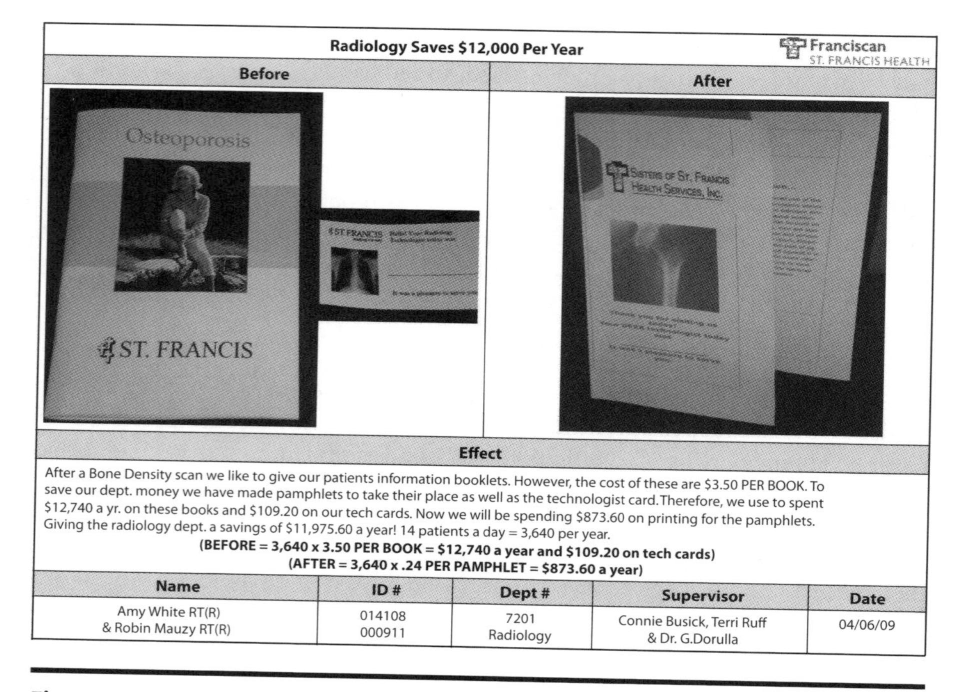

Radiology Saves $12,000 Per Year — Franciscan ST. FRANCIS HEALTH

Before	After
Osteoporosis ST. FRANCIS	Sisters of St. Francis Health Services, Inc.

Effect

After a Bone Density scan we like to give our patients information booklets. However, the cost of these are $3.50 PER BOOK. To save our dept. money we have made pamphlets to take their place as well as the technologist card.Therefore, we use to spent $12,740 a yr. on these books and $109.20 on our tech cards. Now we will be spending $873.60 on printing for the pamphlets. Giving the radiology dept. a savings of $11,975.60 a year! 14 patients a day = 3,640 per year.

(BEFORE = 3,640 x 3.50 PER BOOK = $12,740 a year and $109.20 on tech cards)
(AFTER = 3,640 x .24 PER PAMPHLET = $873.60 a year)

Name	ID #	Dept #	Supervisor	Date
Amy White RT(R) & Robin Mauzy RT(R)	014108 000911	7201 Radiology	Connie Busick, Terri Ruff & Dr. G.Dorulla	04/06/09

Figure 6.2 A Kaizen report shared by the COO.

When you go to gemba, it's an opportunity to sincerely thank everyone involved for their work. The funny thing is, as the leader, I also would like to thank them for allowing me to support them. I can't give direct care, but, contrary to the old style CEO walking around to be seen, it is a privilege for me to see the team. Every time I am in gemba it reminds me of why, just like clinicians, I went into health care. It feels good.[43]

—Kathryn Correia
President and CEO of HealthEast
and former president of Appleton Medical Center and ThedaClark Medical Center

Conclusion

Each level of leader, from a frontline supervisor to the CEO, plays a key role in Kaizen. Leaders at all levels define what leadership is by what they do, especially the senior leaders. There are many key behaviors and mindsets that are the same across levels, such as directly participating in Kaizen. After creating an environment that is conducive to Kaizen, leaders ask for improvements and help create the time required to work on them. Leaders recognize and spread Kaizens, continually reminding people of the purpose and the benefits of their improvements.

Discussion Questions

- What is your answer to an employee asking you "why are we doing Kaizen?"
- Why is it important for senior leaders to directly participate in Kaizen activity, whether that means events or daily Kaizen?
- Which of the senior leader Kaizen behaviors are most commonly practiced in your organization today? Which Kaizen behaviors need to be practiced more frequently?
- What can senior leaders do to directly educate and coach other leaders on the necessary Kaizen behaviors?
- In what ways can senior leaders give recognition for Kaizen?
- How many ideas should you ask each employee for in a year? What is realistic this year? In ten years?

Endnotes

1 Deming, W. Edwards, *Out of the Crisis* (Cambridge MA: MIT CAES Press, 1982), 21.
2 Toussaint, John, and Roger Gerard, *On the Mend: Revolutionizing Healthcare to Save Lives and Transform the Industry* (Cambridge, MA: Lean Enterprise Institute, 2010), 9.
3 Toussaint, John, Unpublished manuscript, 2009, used with permission.
4 Toussaint, manuscript.
5 Toussaint, manuscript.
6 Toussaint, John, email correspondence, June 2011.
7 Toussaint, John, "A Management, Leadership, and Board Road Map to Transforming Care for Patients," *Frontiers of Health Service Management*, March 2013, http://www.createvalue.org/data/blog/Frontiers-29-3%20Spring-FINAL-Toussaint%20exec%20summary.pdf (accessed May 5 2013), 11.
8 Schultz, Rachelle, personal interview, January 2013.
9 ThedaCare Center for Healthcare Value, *Thinking Lean at ThedaCare: Strategy Deployment*, DVD (Appleton, WI: ThedaCare Center for Healthcare Value), 2011.
10 Toussaint, John S., MD, and Leonard L. Berry, PhD, "The Promise of Lean in Healthcare," *Mayo Clinic Proceedings*, January 2013.
11 Graban, Mark, "Podcast #164–Rachelle Schultz, CEO of Winona Health," LeanBlog.org, http://leanblog.org.164 (accessed May 5, 2013).
12 Meredith, Lynne, personal interview, June 2011.
13 Toussaint, John, *Frontiers of Health Service Management*, 7.
14 Brody, Bob, personal interview, January 2013.
15 Jewell, Keith, personal interview, December 2012.
16 Toussaint, John, *Frontiers of Health Service Management*, 9.
17 Jewell, Keith, personal interview, December 2012.
18 Brock, Evangeline, personal interview, July 2011.

19 Tétreault, Michel, personal interview, January 2013.
20 Rudolph, Jennifer, personal interview, September 2011.
21 Graban, Mark, "No Problems Is a Problem—Video," *LeanBlog.org*, May 15, 2010, http://www.leanblog.org/2010/05/no-problems-is-problem-video/ (accessed January 26, 2013).
22 Leonhardt, Kathryn, "Comprehensive Program to Promote 'Fair and Just Principles' Improves Employee Perceptions of How a Health System Responds to Errors," *AHRQ* website, http://www.innovations.ahrq.gov/content.aspx?id=2588 (accessed January 26, 2013).
23 AHRQ, "Patient Safety and Quality: Hospital Incident Reporting Systems Often Miss Physician High-Risk Procedure and Prescribing Errors," *AHRQ* website, http://www.ahrq.gov/research/mar08/0308RA2.htm (accessed January 26, 2013).
24 Jewell, Keith, personal interview, December 2012.
25 Click, Joe, personal interview, July 2011.
26 Mann, David, *Creating a Lean Culture* (New York: Productivity Press, 2005), 92.
27 Toussaint, John, "More Organizational Transformation Topics: Gemba," ThedaCare Center for Healthcare Value (blog), http://www.createhealthcarevalue.com/blog/post/?bid=159 (accessed January 26, 2013).
28 Meredith, Lynne, personal interview, June 2011.
29 Tétreault, personal interview.
30 Toussaint, "A Management, Leadership, and Board Road Map," 14.
31 Tétreault, personal interview.
32 Dague, James O., personal interview, July 2011.
33 Seidelman, Ray, personal interview, July 2011.
34 Blackstone, John, "A Carmaker as a Model for a Hospital?," *CBS Evening News*, June 6, 2009.
35 Grabon, Mark, "Podcast #144–Dr. Dean Grunner, Lean & ACOs at ThedaCare," LeanBlog.org, http://leanblog.org//144 (accessed May 5, 2013).
36 Imai, Masaaki, *KAIZEN: The Key to Japan's Competitive Success* (New York: McGraw-Hill, 1986), 3.
37 Sellers, Bart, personal interview, September 2011.
38 DeBarba, Herb, personal interview, July 2011.
39 Smith, Jennifer, email correspondence, September 2011.
40 Stanfill, Paula, personal interview, March 2011.
41 Liker, Jeffrey, and David Meier, *The Toyota Way Fieldbook* (New York: McGraw-Hill, 2005), 117.
42 Liker and Meier, *The Toyota Way Fieldbook,* 223.
43 Correia, Kathryn, presentation at Lean Healthcare Transformation Summit, June 2012.

Chapter 7

The Role of Other Leaders in Kaizen

Quick Take

- While Kaizen requires the sponsorship and involvement of senior leaders, leaders at all levels need to directly participate in Kaizen in their areas.
- Department managers, or designated *Kaizen coordinators,* take a very active role in facilitating Kaizen changes and managing a departmental effort.
- Leaders at all levels must ask staff to participate in Kaizen and must help create time for this work.
- Supervisors and managers help assess cross-departmental impact and help avoid suboptimization.
- Leaders drive Kaizen success through their actions, encouragement, and recognition of Kaizen activity.

Being "behind" an initiative is quite different than being in front of it.[1]

—Jamie Flinchbaugh
Author and consultant

From Cop to Coach

For Kaizen to be effective, the role of first-level supervisors and managers has to change dramatically, as illustrated by a story from the laboratory at Children's Medical Center Dallas. John Burns, a medical technologist, had worked for

Children's since 1992 and was described by his leaders as, "competent, dependable, technically superb, responsible, and known for helping colleagues," and from John's protests about those labels, you could add humble. Many times over the years, John had been offered a shift supervisor role, but he repeatedly declined. Why was that? John equated being a supervisor with "going to the dark side," because that role represented being "controlling" and "bossy," and under previous leadership, those traits were rewarded. Being a supervisor was akin to being a "police officer" in that setting.

Two years into the lab's Lean and Kaizen journey, John finally accepted that supervisor position. What had changed? The role, along with the expectations for this role from the directors, was more compatible with John's personality and his desire to help others. John said, "The only good thing about having worked under the old management approach was that experiencing that misery burned into my mind how *not* to be as a supervisor." The lab was shifting to a *no blame* culture, with everybody being more focused on processes and systems. There was a renewed focus on their mission of *taking care of kids*, and leaders were expected to act as coaches and helpers, facilitating the success of others. This new approach was palatable to John, as he said, "It's a night and day difference in the culture and I would recommend my position to anyone."[2]

Kaizen Requires Leaders at All Levels

While senior leaders play an important role in establishing a culture of continuous improvement, the managers and leaders who are closer to the work play a critical daily role in encouraging, supporting, and recognizing Kaizen.

Franciscan has found that leadership also comes from those who are not in formal management positions. "There was resistance at the start of the Kaizen program," according to Paula Murphy, the Kaizen coordinator in their pharmacy, as "staff went to the classes but they didn't get it." Then, Paul Shoemaker, a pharmacist, spoke up in a department meeting and said, "This is to enable you, those who know your job better than anyone, to make your job more user friendly."[3] After Paul's comment, Kaizens started rolling in. It takes leaders, from the frontline staff to the CEO, to create the right conditions for Kaizen.

Role of Middle-Level Managers

At Franciscan, middle-level managers include managers and directors. Middle managers play a critical role in the success of Kaizen, because these managers can essentially shut down the Kaizen process if they are not properly engaged. Senior leaders need to understand the roles these managers play in Kaizen so they can provide encouragement and support to these managers while modeling the behaviors they should be emulating.

Paula's Baby Steps Lead the Way

Paula Stanfill's life's passion is caring for babies. Her eyes light up when she talks about the babies she and her staff have cared for. As manager of the Franciscan Neonatal Intensive Care Unit (NICU), Paula admits she was skeptical at first, as she thought Kaizen was another program of the month—just another to-do being pushed on her plate—a plate that was already overflowing and overwhelming at times.

Paula was skeptical at first, saying, "It seemed like improvement was something we already did,"[4] so she wondered why she needed to write up improvements after the fact. It was not until she took some baby steps and completed a Kaizen report of her own that the light bulb went on for her. Traditionally, as manager, Paula thought she was supposed to lead in a directive manner. Instead, she got involved and led the way for her staff by turning in most of the improvements in her department. However, most of her staff initially resisted adopting the change. She explained that, at first, it was like pulling teeth. Like so many things she had brought to them over the years, her staff, being cooperative, said, "Just tell us what to do and we'll do it," but Paula wanted them to motivate themselves.

The "Great Big Pile of Problems"

Paula admits the Kaizen program was time consuming in the beginning. Another manager commented that it was a "great big pile of problems." The piles were created when managers let a backlog of Kaizens build up because they felt the need to oversee and control each one that was submitted. While some Kaizens required coordination between departments (such as maintenance, capital expense approvals), not everything actually requires the manager's direct involvement at all stages.

Early in the program, Paula felt the need to "police" every submitted Kaizen to make sure they met her standard. At one point, she was overwhelmed and decided to put control of the Kaizens back in the employees' hands by asking the lead nurses in her department to get involved. She discovered that they would learn to coordinate their own Kaizens if she asked them to.

Over the next few months, Paula coached her lead nurses as they worked on their own Kaizens. Once those nurses became proficient, she asked them to coach the rest of the staff in her department, as her nurses were well versed in a nursing education practice called "see one, do one, teach one." Beyond lightening some of her own workload, Paula discovered the process grew self-starters in her department and sparked enthusiasm. That's when the program started taking off.

Lynne Meredith, director of revenue management at Franciscan, discovered something similar. After the first year of Kaizen, Lynne and her manager decided to get directly involved only in those Kaizens that required more than one person and especially those that affect her entire team. Lynne said, "I told them you don't need my permission to make your job more efficient."[5]

Paula and Lynne discovered that success comes when leaders at all levels in an organization start by leading the way and then transitioning to more delegation over time by:

1. Learning and modeling the Kaizen practice themselves
2. Coaching their direct reports as they learn to model the practice
3. Asking their direct reports to coach those who report to them as they learn the practice

Key Actions for Middle-Level Managers

Key Action 1: Be the Departmental Owner and Develop Co-Owners or Coordinators

When a Kaizen program is started in a department, the manager should initiate and help coordinate the Kaizen activity. However, about a year into Franciscan's implementation, it became clear that the departmental managers were too busy to adequately manage the Kaizen workload. Franciscan asked each department to appoint a Kaizen coordinator if it was deemed necessary, and a few departments decided to use a small team approach. In addition, some departments that are on a nursing "Magnet" journey now use the *unit council* to review and approve those Kaizens that have departmentwide impact. Kaizen coordinators at Franciscan have *Kaizen program coordination* added to their job responsibilities, because it must be something they are formally expected to oversee.

Role of Departmental Kaizen Coordinators

The pharmacy department at Franciscan's St. Francis Health's Indianapolis campus appointed a departmental Kaizen coordinator. The roles and responsibilities of the departmental Kaizen coordinator are described below. These are the same duties performed by supervisors and managers in the absence of a coordinator.

Sort

Paula Murphy, a Kaizen coordinator at Franciscan, receives between 200 and 300 Kaizens per year from staff in her department. The coordinator role is only 25% of her job, and she does not have time to work with every Kaizeneer, so she sorts Kaizens to decide where to best allocate her time. For the simple ones, she has a quick discussion with the Kaizeneer and has them move forward. For the more complex Kaizens, she helps as much as she can and sometimes directs the Kaizeneer to talk with a person who can help them better, such as their supervisor or the unit manager.

Coach and Coordinate

A Kaizen coordinator takes on part of the supervisor's role of coaching by helping staff members think through the implementation of Kaizens, where appropriate, especially ones that are cross-departmental. From time to time, someone will present a problem to her without an identified solution. Paula either spends the time to help them identify a solution to the problem, or she asks them to work with their supervisor and come back with a solution.

Paula cautions staff that, "Not everything is a Quick and Easy Kaizen." For example, she was presented a Kaizen that asked for a billboard in the break room that lights up and flashes. Paula reminds staff that for it to be a Kaizen, "you do it, you start it, and you complete it." She says, "I'm not Santa Claus."[6]

Occasionally, Paula received a Kaizen that was really an idea for another department to make an improvement. For example, one Kaizeneer wanted to put a doorbell outside the catheterization lab. When she received a Kaizen like this, she helped the Kaizeneer coordinate with other areas. If it was not a complex Kaizen, she would simply give them the contact information and let them follow through and make the contact.

Receive, Approve, and Track; Surface and Share

A Kaizen coordinator takes on the burden of receiving, approving, and tracking Kaizen Reports for the department. Kaizen coordinators also send all approved Kaizen Reports to every department staff member via email or posts them on a bulletin board, so they know what changes are being made.

Key Action 2: Use Departmental Meetings

Managers of the higher-performing departments at Franciscan regularly devote a part of their meeting time to Kaizen. For example, they ask their staff to come to the next meeting with two improvement ideas—small, quick ones. They block off some time in the meeting, such as 15 minutes, to discuss some of those ideas. They ask staff to voice their ideas, and they limit the ensuing discussion to a few minutes for each Kaizen, allocating just enough time to understand the nature of each Kaizen and any issues that would have to be overcome to implement it. These managers record each Kaizen in a simple action item list. In organizations that use the Visual Idea Board method, as mentioned in Chapter 5, these ideas might go onto an Idea Card to be posted on the board, keeping the idea visible for follow-up.

Key Action 3: Encourage Staff to Participate by Asking for Their Ideas

It is the role of managers and supervisors to continuously encourage people to implement their ideas and turn them in as Kaizens. At Franciscan, the revenue management department is a top-performing Kaizen department in terms of *Kaizens per full-time equivalent (FTE)*. Their former director, Debbie Tocco, would periodically visit with each employee to help them think of Kaizens. She would ask if they had changed anything lately. If they had, she would ask what was changed, and often she would find herself telling them, "That's a Kaizen," encouraging them to write it up. If they had not done a Kaizen lately, she would ask them questions about their work. If she saw an opportunity, she would ask a question about it, rather than tell them what the Kaizen should be, in order to help them discover their own Kaizen. She also talked about Kaizens at her department meetings.

Paula Stanfill, the NICU manager at Franciscan, did a similar thing to boost the number of Kaizens being done in her department. She carried blank Kaizen Reports around and asked her staff if they had made any changes lately. If so, she helped them document the changes as Kaizens on the spot. Paula explained that it helped engage her busy staff, and a few individuals had an aha moment when they saw how easy documenting a Kaizen can be. Paula even created her own Kaizen Report, illustrating again that leaders at all levels can participate in their own Kaizens!

Key Action 4: Create a Departmental Recognition System

Middle-level leaders can create a departmental recognition system if the broader organization does not have one, or they can supplement the overall program, as facilitated by the Kaizen Promotion Office. Ways of providing recognition can include department bulletin boards or a "star chart," as used at Franciscan, that has a star posted for each person's completed Kaizen.[7]

Key Action 5: Put a Tracking System in Place, if One Does Not Exist

Design and put into place a mechanism to track Kaizens submitted in your department, if one is not already in place at the organizational level. Methods for tracking Kaizens can include log sheets that show active Kaizens or binders that contain completed Kaizens. Electronic tracking systems have also proven helpful to many organizations, as discussed in Chapter 5.

Key Action 6: Tie to Performance Evaluations

At the right point in time, Kaizen can be tied to your annual employee performance evaluations. In the second year of the program at Franciscan, one department added the expectation of two Kaizens to their performance reviews, and in the third year, several other departments added a Kaizen count expectation to their review process. The experience at Franciscan suggests that it is best to wait until you are a few years into Kaizen before creating this formal connection. Early on, you want Kaizen participation to be voluntary, and even the slightest pressure to meet an annual goal will shift the focus to goals and their compensation instead of focusing on the improvements and learning how to do Kaizen.

The pharmacy at Franciscan had strong Kaizen growth in the first two years, but there were still individuals who would not participate. In their third year, they added Kaizen to their annual performance evaluations in order to encourage those late adopters. However, soon afterward, staff started believing Kaizen was mandatory. Their Kaizen coordinator, Paula Murphy, had to explain to the staff that Kaizen is not mandatory—rather it is encouraged. She explained that staff does not have to do Kaizen. If a person does not turn in six Kaizens per year, they will score a little lower in their performance review, but there are plenty of other categories in which they can score higher and negate the Kaizen score. Paula gets the most pushback regarding Kaizen being part of performance evaluations from staff who previously received the lowest evaluations.

In 2008, Natalie Novak, former director of Franciscan's medical records department, asked her employees to submit one Kaizen per month. She asked her managers to add it to their goals and asked each manager to ask it of their staff. Submitting 12 Kaizens per year became the standard. Her department achieved the goal and became one of the top-performing departments at Franciscan. Lynne Meredith said, "Just saying you have to turn in Kaizens and not holding people accountable doesn't work for 20% of people."[8]

Different Views on Kaizen and Annual Reviews

Although some in the quality field remain adamantly opposed to annual performance reviews,[9] often citing the teachings of Dr. W. Edwards Deming,[10] this process is a reality in most organizations. When these systems are in place, Kaizen organizations often incorporate staff participation into annual ratings or pay increases.

There is a key distinction to make between setting a goal that is aspirational versus having a quota or a hard target that has penalties for not being met. If leaders set a goal of one Kaizen per employee per month, they need to ensure that the overall system can support such a goal, including the capability of supervisors to coach and lead the process and time being made available. If the goal is not reached, top-level managers should follow the common Lean management practice of asking why and looking for barriers to Kaizen, as discussed in Chapter 4. This will lead to more effective development of a Kaizen program than would punishment.

In a fully mature Kaizen environment, employees and managers at all levels will participate in Kaizen because it is personally rewarding, it makes their lives easier, it improves patient care, and it strengthens the organization. With this intrinsic motivation, goals would not be necessary. However, when getting started with Kaizen, goals can be helpful if done with the right approach, including having clear expectations, managers who participate in Kaizen, and fairness. When Franciscan instituted goals, it sent the message that Kaizen is important and should be tried by everybody. There was some grumbling initially, but this quickly subsided once staff members tried Kaizen and saw the benefits.

Jim Adams, former senior director of laboratory operations at Children's Medical Center Dallas, recalibrated their performance evaluation system because, "The scoring had become quite inflated over many years," he observed. Adams set the new baseline of "meeting basic expectations as described in one's job description" as a score of a 2 out of 5 (with 5 being the highest). Adams says there was no guarantee of a better rating for participating in Kaizen or other Lean methods, but it was stated as a basic expectation, and the increase of one's score depended on the value that was added through Kaizen. Placing the expectations in the job descriptions and the performance evaluation tool made it easier to positively differentiate, in a formal way, between those who adopted and operationalized Lean and Kaizen concepts from those who did not.[11]

Role of First-Level Managers

First-level managers are also critical to the success of Kaizen. They are the ones that teach and coach staff about Kaizen on a daily basis. First-level managers supervise frontline staff and have roles and titles such as supervisors, team leads, lead nurses, charge nurses, lead technologists, and patient care coordinators.

Key Actions for First-Level Managers

A first-level manager is the most important person, next to the Kaizeneer, because daily Kaizen is guided and coached most actively by these leaders.

Key Action 1: Coach

At first, Kelly Butler, NICU supervisor at Franciscan, struggled with getting her staff to participate, remarking, "They didn't want to do Kaizen, because it sounded difficult."[12] She helped them by starting simply, including helping staff remove an unnecessary lock from a closet door. Kelly kept encouraging them, but some would say, "I don't know how to do it." So, Kelly would help them think through an improvement, documenting and submitting a Kaizen Report.

Laura Pettigrew, former manager of medical records at Franciscan, recognized the importance of the supervisor in Kaizen success. She involved herself in most Kaizens at the start of her program, with the purpose of modeling the coaching style she wanted to see emulated by her supervisors. She wanted to show her supervisors how to keep the responsibility in the hands of the Kaizeneer and how to lead them through solving their own problems by asking questions and helping them think through a countermeasure. She said, "I wanted to show them how we learn from mistakes and how supervisors could lead staff to work with the other employee to work it out together. Staff used to moan and gripe about someone else and not do something about it. Now they know they have to do something to work it out."[13]

There are times, though, when managers might want to provide feedback or coaching about submitted Idea Cards. This discussion should not be done in a corrective tone, but rather should be a constructive, honest, and respectful dialogue about how to best orient suggestions toward patient needs and make sure we are solving the real root cause of the problem. As with any Kaizen methodology, we want to treat each idea as if it were a gift, being respectful of the person who gave us a so-called *bad idea*.

In the core lab at Children's Medical Center Dallas, between 15% and 20% of submitted ideas are not implemented in the original form, according to John Burns, supervisor of the hematology area. One reason an idea might be rejected is if the change adds more steps to a process instead of reducing waste. Clay York, manager of the core laboratory, explains that some of these early ideas would introduce more waste because employees were not getting to the root cause of the situation. To help counter this, York has focused on providing improved "Lean 101" training to all employees. Even when an idea is not going to be implemented, York "appreciates that a problem was pointed out," adding, "That's what we want—problem solvers; you need people who can see problems."[14]

One example of an idea that the team did "not move forward" with at Children's Medical Center Dallas (they avoid saying "rejected") was a suggestion from a medical technologist to reuse paper printouts from the printers at the laboratory instruments that were a backup system to electronic reporting. The supervisor thought there might be the risk of a mix-up if the paper was reused on the same analyzer. Rather than rejecting the employee's idea outright, the team brainstormed and decided that it would be acceptable to reuse paper on a different instrument that runs different tests, eliminating the risk of reading the wrong test result over the phone.

To help encourage participation in Kaizen, the lab incorporated the understanding and application of Lean principles and tools as part of the basic expectations of all staff. All job descriptions were modified to include the statement, "Follow standard work documents and actively contribute suggestions and process improvements to eliminate waste and add value in alignment with the hospital, section, and department goals and objectives." Jim Adams adds, "We have tried to encourage staff to incorporate the Kaizen thinking into their daily routines."[15]

Key Action 2: Empower Staff—Do Not Do the Kaizen for Them

Some of Kelly Butler's NICU employees required her coaching a few times before they realized they were to think through the Kaizen themselves first and involve others. When it came time to implement, Kelly helped people identify the needed actions. She helped them think through with whom they needed to coordinate, and she helped them think through how to engage others in the Kaizen idea. Then she empowered them to do the actions. She did not do it for them, but instead helped them think through how they could do it on their own. Being a coach means avoiding the temptation to do the Kaizen for people, because that creates dependency (and too much work for the leaders), and it hampers learning.

Key Action 3: Use Rounding to Coach

Cherysh Getz, former supervisor of medical records at Franciscan, visits with every staff member at least once a month and uses some of that time to coach staff on Kaizens. When she sits down with a new staff member, she describes the program and shows them examples. Then, she guides them through submitting a Kaizen Report online. She keeps revisiting with them until they are comfortable creating a Kaizen Report on their own.

To get the brainstorming process going, she asks them questions about how they go about their work and where the items they need to do that work are located. She said, "Once they get started and they realize the benefits of doing Kaizens outweigh the time it takes, they realize they can do it."[16] When Cherysh runs into a staff member who seems resistant, she uses positive encouragement. She realizes there can be a variety of reasons why staff hesitate to document Kaizens. If she discovers they are technologically challenged, she spends extra time guiding them through the online entry until they feel comfortable. Cherysh also teams up with others on every Kaizen she does, because it gives her additional opportunities to coach others.

Key Action 4: Help Set Expectations

The supervisor should also help Kaizeneers set priorities, by helping them identify the most important implementation actions and teaching the Pareto principle that 80% of the benefit will likely come from 20% of the actions. The supervisor coach should also ensure Kaizeneers are not trying to do too much on their own without including the help of others.

When the supervisor comes across a Kaizen that should not be implemented, she should carefully explain why. Ronda Frieje, pharmacy manager at Franciscan, said, "If we couldn't do a Kaizen, I felt that we owed it to the Kaizeneer to sit down with them face to face and explain why."[17] In situations like this, the supervisor should work with the Kaizeneer to find other ideas that could be implemented to address the underlying problem or opportunity.

Key Action 5: Review and Approve Kaizen Reports

The person's direct supervisor is generally the person that approves Kaizen Reports for the Kaizeneer, as introduced in Chapter 5. The direct supervisor's main task is to verify completion of the Kaizen. Did the Kaizeneer do what she claimed to do? Do the results make sense? If so, approve the Kaizen. If not, then work out a plan with the Kaizeneer.

Supervisors Assess Broader Impact

If the Kaizen will impact the entire unit or department, the manager may need to be involved in approving and helping with the Kaizen. Alternatively, organizations seeking or already at Magnet Nursing status may require Kaizens that impact the entire unit to undergo unit council review.

If the Kaizen involves personnel, resources, or processes from other departments, the supervisor may need to help the Kaizeneer communicate and coordinate across departments. Coordinating Kaizens across departments is more challenging than local Kaizens, but the benefits are often greater.

For example, at Franciscan, a nurse worked with the infection control and food services departments to provide disposable towelettes on trays to encourage patients to wash their hands before eating meals.

Concerns about Suboptimization

Leaders at Franciscan had a legitimate concern that staff-driven improvements might not adequately consider system-level effects, leading to potential suboptimization and problems for the larger system. However, after reviewing thousands of Kaizens, the number that had the potential to suboptimize the system was very small. Asking supervisors to coach staff members through their Kaizens helps reduce potential suboptimizations. Also, when Kaizens get discussed prior to implementation, the Kaizeneer often gets pushback from other staff members if it would appear to suboptimize things. Situations like these provide another opportunity for managers to coach and educate their staff members.

In one example, a nurse from an inpatient unit at a hospital identified a problem. Patients who were admitted from the emergency department between 5:00 p.m. and 6:00 p.m. arrived during the shift-to-shift nursing report period. As a result, the patient might not get seen immediately upon admission, which can delay needed medication orders or care.

The suggestion, however, required some discussion among leaders and staff from inpatient units and emergency. Postponing admissions is one countermeasure, and it might seem to solve the problem, as it was stated.

In a traditional suggestion box system, a director or vice president might read this card and reject it, saying that the hospital cannot delay admissions because it causes space and throughput problems in the emergency department. The leader or leaders might just say "no," and the nurse who submitted the card might get zero feedback, with no opportunity to discuss the root cause problem or find another way to address these problems.

In a Kaizen approach, this idea is the beginning of a dialogue. A manager can talk with the nurse, or with a team from the unit, or with a cross-functional team from the unit and emergency. The mindset might shift from talking about "why we can't take patients between 5 and 6" to figuring out "how can we change things so patients can be seen promptly, regardless of their admission time?" This might require process- or staffing-level changes in the inpatient unit, but somebody should be making a decision that is best for the patient and best for the organization rather than only doing what is best for a particular unit.

One outcome of an Idea Card like this might be to initiate a Kaizen Event or a larger analysis and problem-solving effort for what is a relatively complex issue and set of circumstances.

Concerns about Selfish Kaizens

Some leaders also wonder if employees will gravitate toward their own selfish desires when implementing Kaizens instead of serving the needs of the organization.

After reviewing thousands of Kaizens, Franciscan found very few that only helped the implementer without also helping the organization. Even those that one could call *selfish* did, over time, lead to Kaizens that helped the organization. Peer pressure helps to limit selfish Kaizens, especially those that will impact others. The review by supervisors and managers is another important check and balance in this process, and it is one reason why people are not allowed to just make changes in isolation.

Franciscan adopted a practice to overlook those "selfish" Kaizens from new Kaizeneers that have no negative impact on others and to look at the longer-term view of developing staff. This ties into Imai's model of the three stages of Kaizen, where it is important to build enthusiasm in early stages by saying "yes" to most ideas.

Keeping with Imai's model, more mature Kaizeneers might have their ideas challenged in a constructive way, coaching them on their improvement and problem-solving skills, while approaching them with respect and not questioning their intentions. Leaders at Franciscan have learned to carefully consider approaching people who appear to turn in selfish Kaizens and to wait until it becomes an obvious pattern of behavior. Typically, initiating that discussion helped a person come up with a better idea for addressing their problem or opportunity. However, in one case, these discussions with a mature Kaizeneer caused the staff member to temporarily stop submitting Kaizens.

Concerns about Promoting a Quick Fix Solution Mentality?

There is a natural human tendency to start with solutions, especially with best practices like nursing bundles, without first understanding the problem and analyzing the root causes. With Kaizen, implementing an idea should be really easy to do, especially for new Kaizeneers. However, leaders need to develop the Plan-Do-Study-Act (PDSA) and scientific thinking culture in the organization as people grow and mature in their Kaizen abilities, meaning ideas should be tested before they are permanently implemented. Again, at Franciscan, the checks and balances of the supervisor, peer pressure, and management review have seemed to temper this tendency to jump to solutions.

Key Action 6: Help Document Benefits

As Kaizeneers gain more experience, first-level managers can spend more time coaching on how to evaluate the benefits of their improvements, in patient safety, staff safety, time savings, cost savings, quality, speed, patient satisfaction, and staff satisfaction.

Nancy Thompson, a supervisor in the Pediatrics Unit at Franciscan, saw a need to place thermometers at the point of use in each patient room. Her manager realized they would have to be cost justified to leadership, so she spent time teaching Nancy how to create a business case. Nancy collected the data and completed the analysis, which led to leadership approving the purchase. This was not a trivial Kaizen, as it not only developed Nancy's skill of creating a business case, which is knowledge she now uses to coach her frontline nurses, but it also benefited the organization more widely, as discussed later in this chapter.

Key Action 7: Make Kaizen Fun

Kelly Butler helped her employees make their Kaizen Reports more cheerful by adding pictures. According to Kelly, it took staff doing Kaizen to realize, "Oh, this isn't so hard." Over time, her team found it easier to take and incorporate pictures. For example, Kelly asked one of her staff to hold up some old forms and frown to capture the mood before the improvement and then asked the staff to hold up the new colorful forms and smile to capture the mood after the improvement, as shown in Figure 7.1.

Key Action 8: Recognize and Reward

See to it that those employees you coach are properly recognized and rewarded, so they will want to do it again. Ensure they get the credit that is due them. Giving recognition in front of others will encourage more people to participate.

Kelly Butler noted that, when staff got recognized for Kaizen and when they saw that it does make a difference in the daily work in the unit, her entire nursing

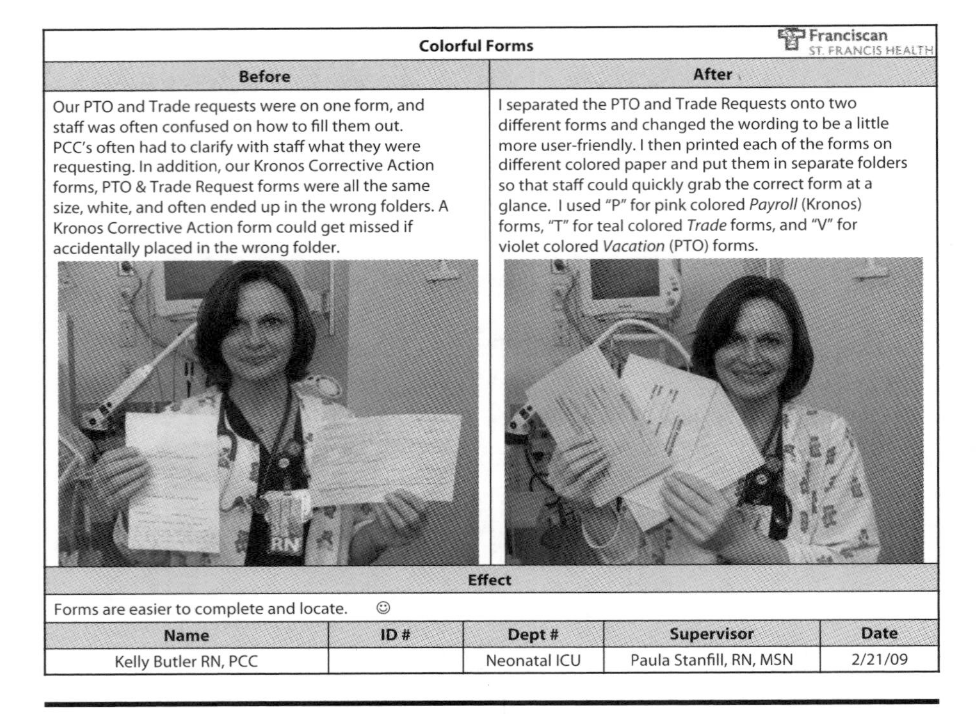

Colorful Forms — Franciscan ST. FRANCIS HEALTH

Before	After
Our PTO and Trade requests were on one form, and staff was often confused on how to fill them out. PCC's often had to clarify with staff what they were requesting. In addition, our Kronos Corrective Action forms, PTO & Trade Request forms were all the same size, white, and often ended up in the wrong folders. A Kronos Corrective Action form could get missed if accidentally placed in the wrong folder.	I separated the PTO and Trade Requests onto two different forms and changed the wording to be a little more user-friendly. I then printed each of the forms on different colored paper and put them in separate folders so that staff could quickly grab the correct form at a glance. I used "P" for pink colored *Payroll* (Kronos) forms, "T" for teal colored *Trade* forms, and "V" for violet colored *Vacation* (PTO) forms.

Effect

Forms are easier to complete and locate. ☺

Name	ID #	Dept #	Supervisor	Date
Kelly Butler RN, PCC		Neonatal ICU	Paula Stanfill, RN, MSN	2/21/09

Figure 7.1 Kaizen reports can be fun.

unit became engaged. Kelly said, "Once we got the ball rolling, people got excited and started thinking up new ideas. I now catch people talking about Kaizens."[18]

Key Action 9: Share and Spread Ideas

The real strategic power of Kaizen is in the sharing and spreading of ideas. An improvement in one area could be the stimulus for staff to customize the idea and apply it in another area. One idea can lead to ten others. Part of the role of a supervisor is to share and spread ideas to others. When you see ideas used in other areas, bring them back and share them with your staff, and then encourage them to implement the ones they like.

After Nancy Thompson placed thermometers at the point of use in all pediatric patient rooms she didn't stop there. Nancy approached the Kaizen Promotion Office (KPO) and willingly shared her idea with the entire hospital system by email, which led to widespread adoption. One idea was multiplied, which reduced the time nurses spent searching for thermometers in many sites. This reduced the nurses' frustration and stress at work and increased the joy in their work. One supervisor really can make a big difference.

Key Action 10: Be a Cheerleader

Finally, first-level managers are the daily cheerleaders of Kaizen. Growing a Kaizen program requires sustained energy and encouragement over a long period of time. The most successful departments at Franciscan have supervisors and managers who continuously encourage Kaizeneers to take action on each Kaizen idea. Even if an idea does not work, they encourage people to reevaluate things and try again. In the face of failure, supervisors remain positive and focus on what can be learned from the failed Kaizen and what can be done better next time.

Leaders Drive Kaizen Success

Kaizen must be led by middle-level managers according to Ronda Frieje, the manager of the Indianapolis pharmacy within Franciscan. One early Kaizen that made a big difference for the pharmacy staff was rearranging the department to improve flow. Everything required to fill an order was arranged in the logical physical order. They had talked as a department about it for a week before making the change. Then, after the change, they tweaked things for several weeks until they found optimal locations for everything. After this improvement, staff saw a clear benefit to them and their work, because things were easier and worked better. Ronda explained that managers "have to help staff with Kaizen for a while until they start getting it. Then, the manager has to start stepping away to give ownership to staff members, coaching and supporting them over time."

Occasionally, employees will quit participating in the Kaizen program. One of Franciscan's most prolific Kaizeneers in 2007 had not submitted anything in 2011. When she was asked why, she said that her new manager did not see the Kaizen program as important. Her previous manager placed Kaizen as a top agenda item, and her department had been a top Kaizen performer. This demonstrates how important leadership is to growing and sustaining Kaizen programs.

Lynne Meredith keeps Kaizen fresh by finding different ways to encourage her staff. She explains, "You can't relax—you have to keep staff engaged by trying new things and rehashing old things in order to keep it interesting."[19] Her leadership team has put flyers in bathroom stalls, conducted departmental contests, posted Kaizens on their bulletin board, and had potluck lunches. Lynne noticed that competition amongst staff helped early in the program. However, as her Kaizeneers matured, they were not as motivated to compete. Unfortunately, competition discouraged the lower performers, because they never seemed to measure up over time, which led them to basically give up. So, in her third year of the program, she decided to shift from competition to cooperation to raise the overall capability of the department. Lynne began having the top performers work with the low performers to get them to see one Kaizen, do one, and teach one. Now, all her staff members are good about sharing and are self-motivated to come up with unique ideas to improve their work and patient care.

Conclusion

Frontline leaders, like John Burns at Children's Medical Center Dallas, play a critical role in the daily practice of Kaizen. They are the ones with the most contact with frontline staff, meaning they have the greatest direct influence on people accepting or rejecting the practice of Kaizen. Leaders at all levels need to not feel threatened by their staff finding problems. Since *no problems is a problem*, the existence of problems should not make leaders feel shame. Likewise, staff members identifying and implementing Kaizens does not mean a leader is not doing their job. Quite the opposite, it means they are creating an environment where everybody is participating in improvement and that should be rewarded. As leaders shift from being a cop to being a coach, everybody wins.

Discussion Questions

- How can you encourage and support your middle and frontline managers to participate more in Kaizen?
- Can you think of a recent change that was made in your organization that was inadvertently suboptimizing? How can the organization shift to a PDSA mindset where a change can be reversed if it caused a problem?
- How can leaders shift from merely rejecting or accepting ideas to being more collaborative and working toward better solutions?
- How can senior leaders evaluate which middle and frontline managers are doing the best job of creating and growing a Kaizen culture?

Endnotes

1 Flinchbaugh, Jamie, *The Hitchhiker's Guide to Lean: Lessons from the Road* (Dearborn, MI: Society of Manufacturing Engineers, 2005), 39.
2 Burns, John, email correspondence, August 2011.
3 Murphy, Paula, personal interview, July 2011.
4 Stanfill, Paula, personal interview, March 2011.
5 Meredith, Lynne, personal interview, June 2011.
6 Murphy, Paula, personal interview, July 2011.
7 Graban, Mark, and Joseph E. Swartz, *Healthcare Kaizen: Engaging Front-Line Staff in Sustainable Continuous Improvements* (New York: Productivity Press, 2012), 257.
8 Meredith, Lynne, personal interview, June 2011.
9 Graban, Mark, "Podcast #117–Samuel A. Culbert, 'Get Rid of the Performance Review!'" LeanBlog.org, http://www.leanblog.org/117 (accessed May 5, 2013.)
10 Deming, W. Edwards, *Out of the Crisis* (Cambridge, MA: MIT CAES Press, 1982), 116.
11 Adams, Jim, email correspondence, September 2011.
12 Butler, Kelly, personal interview, April 2011.
13 Pettigrew, Laura, personal interview, August 2011.

14 York, Clay, personal interview, July 2011.
15 Adams, Jim, email correspondence, September 2011.
16 Getz, Cherysh, personal interview, August 2011.
17 Frieje, Ronda, personal interview, June 2011.
18 Butler, Kelly, personal interview, April 2011.
19 Meredith, Lynne, personal interview, June 2011.

Chapter 8

Organization-Wide Kaizen Programs

Quick Take

- Many organizations follow Kaizen principles in setting up and spreading their Kaizen programs, including starting small (in a single department).
- An organization can start seeing results from a Kaizen program very quickly.
- A Kaizen Promotion Office (KPO) plays an important role in coordinating the program in a larger organization, including training on Kaizen and tabulating overall Kaizen metrics.
- Responsibility for Kaizen cannot be simply delegated to a KPO. Leaders are still responsible for playing an active role in leading and inspiring improvement.
- There are pros and cons of using financial incentives for Kaizen or tying participation to annual reviews.

A journey of a thousand miles must begin with a single step.

—Lao Tzu

From Helplessness to Empowerment

Prior to Kaizen, Paula Stanfill's staff experienced a kind of helplessness. With every new program, her nurses felt more and more overloaded with things to do. Her staff did not understand the Kaizen process until they did a few themselves. Then, Paula enjoyed watching each staff member experience their own aha moments. Slowly, she watched their excitement and enthusiasm grow. It took about one year to get her lead people going. When they realized they could make change in their department, they felt empowered. They, in turn, involved other staff members in Kaizen.

The Kaizen program was the key to a positive change in the NICU. Now, they feel like they can do something about almost anything. This empowerment has brought control back to their work life and a sense of joy back to their department. Paula says, "Now, staff members recognize problems on their own, and they take ownership to team up and resolve them."[1] Instead of the manager leading with a solution, the employees tell Paula about the solutions they will be testing to resolve the problem. Paula has also noticed an increase in people working as a team to identify improvement opportunities and resolve problems.

Paula's department may not have produced the most Kaizens at Franciscan, but the NICU has made a significant impact on the spread of Kaizen. It was the quality of their Kaizens that, when shared, demonstrated to other parts of the hospital the difference Kaizen can make.

From One Department to the Whole Organization

Many of the Kaizen methods shared to this point can be implemented locally at a department level, as did the laboratory at Children's Medical Center Dallas. Other organizations, like Franciscan, have moved beyond local efforts to develop and maintain an organization-wide Kaizen program and culture. This chapter shares what Franciscan and other successful organizations have done to create a programmatic approach to Kaizen. This includes creating a central Kaizen Promotion Office, considering the use of incentives and contests, and documenting and sharing Kaizens electronically. A summary of the costs at Franciscan can be found in Table 8.1.

Getting Started

The upfront costs can be as small as providing some initial kickoff training. As mentioned earlier in the book, Franciscan's leaders brought in Norm Bodek for a day of training and workshops. If you bring in a well-known consultant or a trainer from a local community college, the costs and fees can vary widely. You can try to create your own kickoff training, but the expertise brought by consultants can be invaluable, especially when you are bringing something new to the organization. Yet the expertise level also varies widely, and does not necessarily correlate with the cost, so shop wisely.

A timeline for an organization-wide Kaizen program includes planning, kickoff, growing, and maturing the program. At Franciscan, leaders planned for a month prior to the official kickoff in April of 2007. The kickoff event lasted for one day. In the two months after kickoff, they provided the initial *how to* training and developed an internal website with blank Kaizen Reports, instructions, and resources. Then, starting with month three, they conducted training on Lean topics

Table 8.1 Summary of Potential Costs to Start an Organization-Wide Kaizen Program

Item	*Cost*
Initial kickoff training	Confidential
Bulletin boards	$4,000 one-time cost
VIP points incentives	$4,000 per year
Contests	$5,000 per year
Electronic Kaizen system	$20,000 initial cost plus ongoing maintenance
Printing forms	$400 per year
Departmental lunch awards	$1,000 per year
Plaques and catering for annual ceremony	$2,000 per year
Staff time and KPO time	Confidential

on a monthly basis for the first year.[2] In the second and third year, Franciscan's leadership brought additional training to specific departments that were struggling with Kaizen. A successful Kaizen program does not just happen—it needs to be designed, implemented, and executed well.

Starting Small and Spreading Kaizen

Just as Kaizen work in a department might start with baby steps and small improvements, the same can be true for a hospital or health system. Thinking back to psychologist Robert Maurer's work, introduced in Chapter 2, it might seem scary to introduce a Kaizen program throughout the entire organization all at once. To help get change started, Maurer asks his clients what is the "smallest step possible" that could improve the organization?[3]

Many organizations take that smallest step by introducing Kaizen methods in a single department, taking the opportunity to learn from these early experiences. The broader organization can learn lessons including:

- How much leadership involvement is required at each level, locally and at director or vice president levels?
- Which methods work best for facilitating and documenting Kaizen changes?
- What is the best way to engage frontline staff to motivate them to participate in Kaizen?
- What recognition or rewards are needed to encourage Kaizen?

By starting small, best practices and lessons learned can be shared as Kaizen methods spread throughout the organization. Additionally, any early mistakes have a limited impact, meaning the organization is more likely to recover from any early missteps. When designing and spreading a Kaizen program, use Kaizen and Plan-Do-Study-Act (PDSA) principles to increase your chances of success.

At St. Elisabeth Hospital, they started with an improvement board in one department, which Mark saw during a visit in 2009. Three years later, they had boards in 75 departments through a process of "self-spread," meaning boards were not forced or pushed on departments. Departments chose to start using the boards as the result of a *learning environment* that was created, including weekly collaboratives where good practices, like the boards, were shared with other departments.[4] The boards, along with some other Lean practices, are "obligatory" in all departments, but "it never would have worked to make it obligatory before the self-spread—they would have hung up the boards, but they would have gone unused," said Marc Rouppe van der Voort, innovation manager at the hospital.[5]

When good things are happening and people see value in early Kaizen efforts, they will want the process to be introduced into their areas. After one department pilots Kaizen, it's common to have multiple department managers or directors actively vying to be next.

When Will You See Results?

Leaders often ask how fast they can expect to see results after launching a Kaizen program. The good news is that Kaizen delivers small, but real and tangible, results immediately. You will start seeing small results as soon as the Kaizens begin to roll in. At Franciscan, the first Kaizen was completed and turned in the very first day after the kickoff event, with dozens more being done in the first month. Each Kaizen provides benefits, even if they are as simple as saving two seconds of time.

It is important to understand that the greatest benefits of Kaizen are seen when the organization's culture is changed. The time required to change a culture is measured not in weeks or months, but in years or even a decade, depending on how aggressively and skillfully Kaizen is introduced and embraced. At some point, Kaizen will permeate the organization so fully that the effects cross over into every improvement activity, making it somewhat difficult to separate the effect of Kaizens from Lean Six Sigma and other improvement approaches that are taking place simultaneously.

James Dague, retired CEO of IU Health Goshen Hospital, recalls, "The first three years of this program are very tough because everybody's waiting for it to go away. You're not going to take all the negatives that got you to this point, where you need an improvement program, and wash it out of your organization in one year." Leaders need to have patience, as Dague adds, "You have to just keep beating the drum. Sometimes you may feel like you're in a stadium all by yourself, but you've got to just persevere with the program. It has to have a return as you go along. But, you've got to establish that this is not a project of the month or a temporary thing, but this is the way we're going to run our culture from now on."[6]

Since the time Franciscan started their Kaizen program, it has spread through the organization each year, leading to some visible changes in the culture. For example, four years ago employees tended to regularly voice their frustrations in a loud and adamant way to administration regularly in various forums around the hospitals. Now, there is noticeably much less of that. The same forums still exist, and employees are encouraged to voice their frustrations and needs, but staff now realize they have the power to improve and redesign their workspace, and the future of the hospital is in each employee's hands, more than ever.

Senior leaders are generally the most interested in results and how Kaizen benefits the organization financially, because they are held accountable in that regard by the board of directors. Although the biggest benefits of Kaizen cannot easily be measured, financial benefits are something that can be objectively measured. According to Keith Jewell, Franciscan COO, "we took 32 million dollars of bottom-line costs out of our three hospital system in Indianapolis in the last few years."[7] More than 12% of that (about $4 million over time) was achieved through their Kaizen program, another 20% of that resulted from the Lean Six Sigma program, and the remainder came from their supply chain value analysis initiatives. Jewell foresees the Kaizen program exceeding other programs in the future, because it engages all staff.

Tying Kaizen to the Organization's Strategy

One key method to help foster a culture change is tying the Kaizen program to the organizational strategy. Early in a program, you want Kaizen to be primarily self-directed, as ideas are surfaced and driven by staff, focused on things that are in their scope of control that matter to them. However, as Kaizeneers develop their skills and abilities, they will naturally enlarge their scope of Kaizens, and we will expect a greater portion to align with what is important to the organization. This alignment is most likely to happen after putting a formal mechanism and communication in place that ensures staff members know what the organization's vision and strategy are. This can accelerate the maturation of the Kaizen process and can ensure that a higher proportion of Kaizens are aligned to higher-level strategy.

At IU Health Goshen Hospital, retired CEO James Dague put a lot of effort into constantly communicating the four *focus areas* for the organization, what some might call a "true north"—customer satisfaction, quality, cost effectiveness, and best people—as well as the overall mission. This helps ensure that their staff-driven improvements naturally align with those strategic areas, in addition to making their own work easier.

Dague says, "We've got pressure through the culture of the organization to always be improving. So you implement 4,000 ideas in a year. Some of those are directed by projects from administration, flowing off of our goals, while you also get idea generation from people simply finding different ways to improve their job." Dague wants employees to be strategic, but he also tells them, "If you've got a job irritant, get rid of it. You can get rid of it through our cultural processes. Pick one and let's get at it, but don't just ignore it."[8]

It is important to help align Kaizen to patient needs and organizational goals. In conjunction with the Visual Idea Board, the laboratory at Children's Medical Center Dallas started tracking and posting key performance measures every day in the lab, near their board. They started holding daily team huddles in that space. Jim Adams, former senior director of laboratory operations, recalls that, within a month of starting to discuss patient-focused measures like turnaround time, "the ideas being generated shifted from 'here's what I want,' like providing a toaster or hot chocolate in the break room, to 'here's what we need to provide better patient care,' such as installing bar code scanners to help reduce clerical errors."[9]

Some organizations issue specific *Kaizen challenges* to staff, where leaders establish and communicate a particular goal or focus area for a period of time, like 30 days.[10] A challenge might focus people on things like improving hand hygiene to reduce infections or improving the emergency department registration process to get patients back to an exam room faster. Generally, these challenges can be applied to other strategic pillars, such as cost or patient satisfaction. After that challenge period, a single improvement could be selected for a prize or special recognition, but all improvement ideas are evaluated and potentially implemented, with recognition being given through the usual methods. Challenges can help align an organization by ensuring improvement is both top down and bottom up.

The Kaizen Promotion Office

Many organizations have created a central Kaizen Promotion Office (KPO) that oversees the program for the hospital, a region, or a system. The KPO is vital to the program's success because it helps guide and coach the organization, while helping to create cohesiveness in the organization's approach to Kaizen.

It is a small distinction, but it is important to note that a KPO does not *own* the program—it is owned by the organization and its leaders. Leaders are still responsible for Kaizen and creating a culture of continuous improvement, from the CEO and

COO on down. A KPO *facilitates* details of the program on behalf of the organization and for the best interests of all stakeholders. A KPO often reviews ideas after they have been tested and implemented, providing feedback and coaching to the supervisors who were involved in specific Kaizens. A KPO also generally plays a role, as does Franciscan, in awarding points to staff members that can be redeemed for products.

Franciscan's KPO has helped other hospitals launch Kaizen programs with varying success. A common denominator of those that performed poorly was that they tried to run a program without a KPO because they did not want to allocate a budget. The problem is, it is not respectful to the staff to ask already busy people to oversee the Kaizen program on top of everything else they were already doing. It is also likely a waste of time to attempt to implement and run a Kaizen program with a lackadaisical effort. It ultimately leads to frustration and then the incorrect conclusion that Kaizen has failed versus the implementation has failed.

Staffing the KPO

The most costly investment in Kaizen is in human resources—people's time and salaries. A few critical human resource items that will be helpful are creating a KPO, selecting a champion, and dedicating a person or people to facilitate the program's activities.

Since Kaizen involves everyone in an organization, a program should consider how to keep everyone energized over time. The energy required to sustain any type of program is derived from the strength of the purpose of the program and the people who facilitate it. Be sure to ensure that KPO staff members are passionate about empowering all employees to do Kaizen and about creating the kind of culture that is described in Chapter 4. Franciscan's KPO found that a team approach was helpful, as they bounce ideas off each other and consider options from more dimensions.

At Franciscan, the COO is the champion. When the three Franciscan St. Francis Hospitals in Indianapolis launched their Kaizen program, they had 25% of two people's time to facilitate the program for a total of 0.5 full-time equivalents (FTEs). One of those people was coauthor Joe, and the other was Mischelle Frank, who, as a nurse, helped connect the program with the nursing staff. As the volume of Kaizens has grown, Franciscan increased their staff support. They now have one regional Kaizen coordinator, who dedicates 50% of her time to Kaizen, along with 25% of the manager of business transformation (Mischelle) and 25% of the director of business transformation (Joe), for a total of 1.0 FTEs, focused on running the Kaizen program for their 4,000 employees.

Similarly, Baptist Health Care in Florida has one coordinator for their Bright Ideas program who dedicates 50% of her time for their 6,000 employees.[11] However, their program is mature and well honed. We recommend that resources be front-loaded during the early phases of a Kaizen program. Then, as the program matures and activities are systematized, the central resources can be adjusted accordingly as the Kaizen management practices are embedded in daily management activities.

Activities of the Kaizen Promotion Office

Franciscan's Kaizen Promotion Office does the following to support their program, each of which will be described in more detail in this chapter:

1. Facilitates the practice of Kaizen
2. Reports Kaizen metrics
3. Coordinates recognition and rewards
4. Enables Kaizen sharing across the organization
5. Develops Kaizen standardized work
6. Develops and delivers staff education
7. Facilitates the documentation and tracking of Kaizens

Activity 1: Facilitates the Practice of Kaizen

Franciscan's KPO helps the organization facilitate the practice of Kaizen. The KPO periodically interviews staff members and leadership, searching for input that will improve the practice of Kaizen in the organization. A study at Franciscan, in late 2010, showed that the most effective drivers of Kaizens in our top-performing departments are supervisor coaching, education, and recognition. Therefore, the KPO focused their efforts in 2011 on encouraging these drivers in departments that were struggling. The KPO surveyed staff members who had not yet participated in Kaizen and found that over half of them believed, incorrectly, that a Kaizen must be a substantial improvement. Therefore, the KPO improved its communication to remind staff to practice Kaizen by starting really small.

Activity 2: Reports Kaizen Metrics

Franciscan's KPO is responsible for tracking and reporting monthly metrics for departments and the broader organization, including the participation rate and absolute numbers of generated Kaizens. For instance in 2012, 36% of staff and 75% of departments participated in the Kaizen program within Franciscan. There were 4,019 Kaizens completed at the three Franciscan hospitals, which is just under one Kaizen per employee including part-time employees, or 1.4 per FTE. The 2013 goal is 5,000 Kaizens. Franciscan's leaders believe there is still a huge opportunity to continue to expand the use of Kaizen in their organization.

The KPO provides top-level managers with a detailed accounting of the Kaizen program's participation rate and financial impact, which serves to remind them that the benefits far outweigh the costs. Department managers can decide if they would like to print and post their numbers or their performance compared to other departments on their communication boards.

At Franciscan, the completed Kaizens in 2010 resulted in a total documented savings of over $3 million. About $1.7 million of that savings was dollars that flowed directly to the bottom line, and over $1.4 million of that was *potential* dollar savings through, for example, the saving of someone's time. They try to clearly distinguish how *potential* savings do not immediately flow to the bottom line of the organization's income statement. For example, saving one hour of a nurse's time may not mean one less hour of wages paid. Instead, that time may be used for another purpose such as enhanced patient care, which can be difficult to put a dollar number on. Franciscan recognizes that their cost savings dipped in 2011 and again in 2012 as they de-emphasized a cost reduction focus in their Kaizen program. In 2013 Franciscan plans to rejuvenate a healthy cost savings focus, and the healthcare industry pressures to reduce costs will help support that focus. Franciscan's program shifts over time to support the strategic focus of the healthcare system, but maintains a strong focus on engaging all employees in making their work processes better for everyone.

Franciscan's KPO tallies dollars saved in the Kaizen program because even a nonprofit or government enterprise requires a positive return to have money to reinvest in the future of the organization or to repay loans that financed construction. In a Kaizen approach, employees should consider the financial impact of ideas. It is not that finances are the most important thing, but they are one of the many considerations when we carve out time in our day to make improvements.

The participation rate and number of Kaizens are objective numbers that are easy to agree on, but savings can be more subjective. In order to ensure the robustness and accuracy of the financial impact data, Franciscan's KPO assists the Kaizeneer in detailing and verifying the financial documentation for any potential savings identified as being more than $1,000. For those that end up being documented as saving more than $20,000, a member of the financial department reviews them for calculation accuracy and evidence of potential or bottom-line impact.

Like many hospitals, Franciscan leadership holds a monthly or bimonthly management meeting. At one of those meetings each quarter, there are 15 minutes reserved for reporting the status of the Kaizen program to all managers, directors, and vice presidents. Participation is reported to the organization by regularly publishing the following Pareto charts:

- The top 25 departments by year-to-date Kaizen count
- The top 25 departments by number of Kaizens per full-time employee

At Franciscan, leaders are careful to focus on the positive by continually recognizing high performers. They are careful not to harm the dignity of the low performers by publishing anything with their names on it.

Activity 3: Coordinates Rewards and Recognition

> Kaizen is an important way we visibly recognize and reward employee initiative.
>
> **—Mischelle Frank**
> *Franciscan nurse*

At Franciscan, the KPO is responsible to pull the data on a monthly basis about to whom VIP points are to be awarded, and sends it to the human resources department to ensure employees receive their proper points.

Additionally, the Franciscan KPO conducts an annual awards ceremony in the auditorium. Kaizeneers are invited who met the annual individual goal, along with representatives from departments that met their departmental goals and the Kaizen high performers.

Franciscan's leaders hand out award plaques to their seven best-performing departments in Kaizen count, their seven best departments by Kaizens per FTE, the top seven money-saving Kaizens, and the seven individuals with the highest Kaizen count. Everyone who met the annual individual goal gets a certificate.

At Franciscan, there are several quarterly contests that reward managers and staff for participation in the Kaizen program. Winners are drawn randomly from a hat, with no analysis of the merit of each Kaizen. For example, they conducted contests that rewarded the three top departments for having the highest staff participation percentage and the highest *increase* in their staff participation percentage with a party. Franciscan also had a contest that rewarded three first-time Kaizen participants, one with an iPod Touch, one with an iPod Shuffle, and one with an iPad. Each contest is different and is tailored to what Franciscan is trying to influence and grow in the Kaizen program at the time. Franciscan budgets $5,000 for Kaizen contests annually.

Activity 4: Facilitates Kaizen Sharing across the Organization

Franciscan's KPO helps ensure Kaizens are posted in prominent places for all to see—cafeterias, departmental bulletin boards, and so on. Franciscan also recommends that each campus have a Kaizen bulletin board that is updated periodically. They found that updating the theme of the board to correspond to the season or a holiday makes it obvious the board has been updated.

Activity 5: Develops Kaizen Standardized Work

The Franciscan KPO is responsible for developing and documenting the standardized work associated with the Kaizen program, such as step-by-step process documentation and supporting educational materials. This is done by discovering, documenting, and sharing Kaizen best practices throughout an organization. For example, to facilitate Kaizen reporting, the KPO purchased digital cameras for each campus and created simple instructions for checkout and use. Kaizen educational materials and process documentation are readily available from the Franciscan intranet.

Activity 6: Develops and Delivers Staff Education

The Franciscan KPO created *how to* training materials for managers to use to train their staff on what Kaizen is and how to create a Kaizen Report. They also offered to conduct training for their staff. About one third of the managers accepted the offer in the first 2 months of the program. The remainder of the managers either provided the training on their own or did not conduct the training. Brief Kaizen training is provided for 5 to 10 minutes to all new employees during orientation. Kaizen is also included in the annual mandatory training for all staff.

Starting in late 2007, Franciscan offered a series of optional 60-minute sessions to all staff on a number of specific Lean topics. The training was taken to each campus, being offered enough times throughout the day and various days of the week to make it convenient for all shifts and weekend option staff.

In another example, the Utah North Regional of Intermountain Healthcare has developed a 60-minute training session for all managers, directors, and senior leaders that covers topics including employee engagement, the role of leaders in managing improvement, how to promote ideas, and coaching staff members through improvement ideas.[12]

Activity 7: Facilitates the Documentation and Tracking of Kaizens

Franciscan's KPO developed and facilitates the methods for documenting and tracking of Kaizens in a home-grown electronic system, as described more fully in *Healthcare Kaizen*.[13]

Besides these seven activities, arguably the most important role of a KPO is to continuously infect an organization with the philosophy of Kaizen. Mischelle Frank summarizes their efforts by saying, "Empowerment is just a nice concept unless there is a mechanism and a support system to enable staff to develop their problem-solving and continuous improvement skills and abilities."[14] Franciscan has found that an effective KPO provides a support system for staff and for leaders at all levels.

Sustaining a Kaizen Program: Incentives and Rewards

During the second year of their program, Franciscan introduced an incentive system that nominally rewarded people with points for turning in completed Kaizen Reports. All Kaizen Reports are treated equally, as there is no ranking of ideas based on their size. Kaizens with large cost savings are treated the same as a Kaizen about moving the location of a stapler. They have no direct cash payouts for ideas, and they reward for the process of doing a Kaizen.

Franciscan provides staff with a small incentive of 200 *VIP points* for each Kaizen that is completed and approved. VIP points can be used to purchase merchandise or gift cards. The points are given to each Kaizeneer listed on the Kaizen. The only qualifier for the incentive is that the Kaizen is documented and approved by the supervisor.

At Franciscan, the KPO also gives out rewards when teaching Kaizen or when presenting at a department meeting. They use rewards, like a $5 cafeteria or coffee shop gift card, Post-it® brand note pads, or small writing notebooks, when staff document a Kaizen Report in the training session based on something they have done in the past but had not yet documented. These come after a short training session and are like Kaizen documenting parties.

At the start of their program, the incentive system cost a total of about a $1,000 per year. Four years into the program, the annual Kaizen costs for the incentive system are approaching $4,000. The human resources department picks up the costs for the overall incentive program, as it was implemented for other incentive purposes, such as thank you card incentives, recognition, and birthday wishes.

Park Nicollet Health Services gives their employees *Ovation Points*, which can be used to purchase items. Jennifer Rudolph, former Kaizen Everyday Engagement Program (KEEP)[15] administrator and lead specialist, says the points "might pique someone's interest to take initiative and learn about Kaizen, but it's not really a big motivator to participate." Park Nicollet gives small rewards for most ideas, but might reward $150 for a large idea that saved $10,000, says Rudolph. She adds that people aren't using the system just to get points. Rudolph says a points system is "nice to have," but not a critical factor.[16]

Pros and Cons of Financial Incentives

Purely monetary rewards can be a disincentive and tend to distort people's motivation for Kaizen.

Organizations in other industries have sometimes learned that financial incentives for Kaizen can backfire if the focus for people shifts from improvement to the reward itself.[17] In some organizations, employees were known to submit ideas solely for the reward payout. In some of these cases, a reward was given for a mere suggestion, as opposed to something that was actually implemented. To help avoid these dysfunctions, keep these guidelines in mind for a rewards program:

- Incentives should be relatively small.
- Incentives should be based on the implementation of an idea.
- Incentives can be paid for attempts at implementation that were ultimately not accepted as a change (following the PDSA approach).

Ultimately, your organization needs to decide the proper role of incentives and rewards based on your own culture. Some pros and cons of incentive payments are listed in Table 8.2.

The Utah North Region of Intermountain Healthcare does not use official rewards or incentives for their idea program. Some units might give meal tickets or candy bars for ideas, but the primary recognition is given verbally in huddles and staff meetings.[19]

Table 8.2 Pros and Cons of Incentive Payments

Pro	*Con*
It is one form of recognition	Focus might turn to earning money rather than fixing the process
Can garner participation from some who would not participate otherwise	Adds another administrative layer to the Kaizen process
Might seem fair to share savings and benefit with staff	Can be hard to gauge the benefits of certain ideas
	Some research shows that financial incentives hamper creativity[8]

Conclusion

If you would like to start an organization-wide Kaizen program, it will experience better results if the program is developed well and is committed to by top-level managers. Beyond their direct participation and leadership of Kaizen, senior leaders must commit adequate financial resources, including staff and leader time.

The organization must also put a Kaizen Promotion Office in place that is staffed with dedicated people who are capable of driving such a program with purpose and energy. A timeline will need to be developed to plan, kick off, grow, and mature a program. The KPO that is put in place needs to facilitate the practice, report metrics, coordinate recognition and rewards, enable sharing, develop standard work, develop and deliver education, and facilitate documentation and tracking. Systems for electronic management and retrieval of Kaizen ideas should be strongly considered at some point in the development of the program.

Discussion Questions

- What are some of the advantages of starting small with a Kaizen program rather than starting with an organization wide program?
- At what point in your organization's Kaizen journey should you consider creating a Kaizen Promotion Office? Can these duties be incorporated into an existing department?
- What would you like your senior leaders and the organization to do to recognize Kaizen accomplishments? What are some fun ways to do this?
- What is the expected or needed role of financial rewards for Kaizen activity in your organization? What are some of the possible downsides or dysfunctions of monetary incentives?

Endnotes

1 Stanfill, Paula, personal interview, March 2011.
2 Graban, Mark, and Joseph E. Swartz, *Healthcare Kaizen: Engaging Front-Line Staff in Sustainable Continuous Improvements* (New York: Productivity Press, 2012), 293.
3 Maurer, Robert, *The Spirit of Kaizen* (New York: McGraw-Hill, 2012), 12.
4 Graban, Mark, "Lean Healthcare Transformation Summit 2012, Day 1," http://www.leanblog.org/2012/06/lean-healthcare-transformation-summit-2012-day-1/ (accessed January 30, 2013).
5 Rouppe van der Voort, Marc, personal interview, January 2013.
6 Dague, James, personal interview, July 2011.
7 Jewell, Keith, personal interview, December 2012.
8 Dague, personal interview.
9 Adams, Jim, personal interview, April 2011.
10 KaiNexus, "Dr. Julie Lewis—on KaiNexus Challenges," YouTube Video, http://www.youtube.com/watch?v=C7GW0QJDr-Y (accessed January 6, 2013).
11 Brophy, Andy, and John Bicheno, *Innovative Lean* (Buckingham, England: PICSIE Books, 2010), 138.
12 Sellers, Bart, personal interview, September 2011.
13 Graban, Mark and Joseph E. Swartz, *Healthcare Kaizen: Engaging Front-Line Staff in Sustainale Continuous Improvements* (New York: Productivity Press, 2012), 281.
14 Frank, Mischelle, personal interview, October 2011.
15 Graban and Swartz, *Healthcare Kaizen*, 160.
16 Rudolph, Jennifer, personal interview, August 2011.
17 Pink, Daniel H., *Drive: The Surprising Truth about What Motivates Us* (New York, Penguin, 2011), 35.
18 Pink, Daniel H., *Drive: The Surprising Truth about What Motivates Us.* (New York: Penguin, 2011), 35.
19 Sellers, Bart, personal interview, September 2011.

Chapter 9

Conclusion

> Better is possible. It does not take genius. It takes diligence, moral clarity, ingenuity. And above all, it takes a willingness to try.[1]
>
> — **Atul Gawande, MD**

Small Methods Lead to a Meaningful Impact

Have you seen *The Karate Kid*? In the movie, Daniel, a teenager who was beaten up by classmates in his new city, started training under Mr. Miyagi, a karate master. Daniel's first assignment was to wash Miyagi's cars. Miyagi showed him how to wax the cars with a particular arm motion— "wax on, wax off." Daniel also had to sand Miyagi's backyard deck with a "big circle" motion. Finally, Daniel had to paint Miyagi's wooden fence and house with specific "up, down" and "side, side" motions.

At this point, Daniel thought he was being was taking advantage of for selfish home improvement needs. How could these chores be helping him train for karate? Daniel was ready to give up, but Miyagi stopped him and said, "Show me wax on, wax off." As Daniel showed him the technique, Miyagi threw a punch toward him and the waxing motion blocked the punch. Daniel's eyes opened wide. He realized that the motions from the chores were really basic karate skills.

Tools and Philosophies

Initially, Daniel wanted to learn karate so he could defend himself or exact revenge on the bullies who tormented him. Daniel was facing a short-term crisis, which meant he was not very interested in exploring the deeper philosophies of the martial arts.

Leaders facing a crisis, in any industry, sometimes want to quickly learn tools, tips, and tricks to solve today's crisis. In doing so, they sometimes miss the opportunity to study, understand, and practice the deeper mindsets of Kaizen that might serve them better over the long run. Your organization can start with the tactical methods, such as Visual Idea Boards, introduced in Chapter 5 (and described in much greater detail in Chapters 5 through 7 of *Healthcare Kaizen: Engaging Front-Line Staff in Sustainable Continuous Improvements*), but sustained usage and success will depend on a sustained and consistent change in leadership behaviors—starting at the top.

Building the Culture

In his introduction to this book, Gary Kaplan MD, CEO Virginia Mason Medical Center said, "The value of using Kaizen to improve health care systems is indisputable." His system has learned that, "in the right culture, Everyday Lean Ideas can flourish—but getting there is not a simple matter." Kaplan said, "Learning to use Kaizen consistently and effectively requires serious culture change and takes many years. This is really not unexpected, as using Kaizen requires deep organizational changes—changes that challenge long-held beliefs and many accepted practices." Virginia Mason started their Kaizen program in small pilot groups back in 2005 and 2006. The system then slowly and carefully rolled it out throughout the organization over time.[2]

Jennifer Phillips, innovation director in Virginia Mason's Kaizen Promotion Office, has been helping build their "overall Kaizen culture" by teaching the philosophy and how to identify waste. Phillips says the system, even with the committed and active leadership of Kaplan, does not yet have "the level of organizational traction we would like. We'd like to see more widespread participation." It takes time in a different way, meaning people must make time each and every week (or day) for Kaizen. This transformation takes prolonged effort over time, meaning that it won't happen overnight. Phillips says clinical areas struggle, "where clinical staff say they like the concept but lack the time to take on such projects." Remember, as we discussed in Chapter 4, a lack of time might be the very first problem we need to solve through Kaizen practices.

Over the past seven years of effort, the Kaizen program, Everyday Lean Ideas, has led to "literally thousands of simple staff ideas [that] have given Virginia Mason tremendous momentum in its mission to eliminate waste and add value for patients."[3]

> Most companies don't die because they are wrong; they die because they don't commit themselves. They fritter away their momentum and their valuable resources while attempting to make a decision. The greatest danger is standing still.[4]
>
> **—Andy Grove**
> *Former CEO of Intel Corporation*

A Minute to Learn, a Lifetime to Master

Daniel could not become a karate black belt overnight, nor can an organization create a Kaizen culture that quickly. Kaizen is about developing and growing basic improvement skills in everyone, every day. There is no better day to start than today. Kaizeneers and leaders need to practice the Plan-Do-Study-Act (PDSA) mindset continually, over many years, to create a lasting culture of continuous improvement. This culture will contribute greatly to meeting patient needs, creating a more enriching workplace, and having an organization that is financially secure for the long term.

> If you want one year of prosperity, grow seeds. If you want ten years of prosperity, grow trees. If you want 100 years of prosperity, grow people.[5]
>
> **—Chinese Proverb**

The basic concepts of Kaizen might seem simple at first. Start your participation by implementing a Kaizen of your own. Then, ask your employees for ideas. Say yes to most of them and work together to find workable alternatives to the so-called bad ideas. Let people implement their own ideas, but help them as a servant leader, when needed. Document the improvements simply. Recognize and thank people for their improvements. Share the ideas with others.

The term *Quick and Easy Kaizen* refers to employees identifying and implementing easy improvements that can be done quickly. Creating, growing, nurturing, and sustaining a Kaizen program is neither quick nor easy in a department or a healthcare system. Leaders need to help initiate and support Kaizen, while working tirelessly to create the conditions that encourage people to openly identify problems and work together with their colleagues on improvement. Kaizen requires leaders at all levels to actively make time to inspire, coach, mentor, and recognize people.

> *It's not what you do once in a while. It's what you do day in and day out that makes the difference.*
>
> **—Jenny Craig**

It will take time to create a Kaizen culture. Through this process, leaders and frontline staff will challenge themselves to grow, personally and professionally. It is inevitable that there will be missteps and lessons learned along the way. It is important not to let small setbacks discourage your continued learning and progress, individually or as an organization.

Building upon Franciscan's Success

As described in this book, the Franciscan St. Francis Health System has made great strides in building a culture of Kaizen and continuous improvement. Starting small in 2007, over 4,000 individuals in the hospitals have generated over 17,000 ideas with an estimated hard dollar savings of over $4,700,000. In 2013, Franciscan plans to re-engage their staff in the simple pleasures of making things better.

CEO Bob Brody summarizes Franciscan's goals and efforts:

> Kaizen is a program our employees embrace to the benefit to our patients, the organization, and our community. In Kaizen, we want everyone's best ideas on how to improve their job, the patient experience, and the overall organizational performance, one small step at a time. Pervasive in the spirit of Kaizen is the desire to always do better, and everyone in the organization can contribute to that. To make that happen, we have to have a mechanism in place to both solicit and to act on Kaizens. It requires an infrastructure, and it is well worth the investment, not only financially, but also from a morale standpoint. People take pride in their ability to contribute, and when you can recognize them for their contributions it just reinforces that positive behavior. The reason we do so well is that we have so many smart people who want to contribute, and if you get out of their way, good things happen.[6]

We hope you will learn from Franciscan, without feeling the need to directly copy everything they have done. That wouldn't be in keeping with the spirit of Kaizen. Adopt the practices that make sense, test through the PDSA process, and adapt them to your needs.

Your Next Steps

In this book, we have shared core Kaizen mindsets, as well as different methods and approaches for managing local improvements and broader programs. While we have shared some recommendations and lessons, we believe strongly that there is no single right way to do Kaizen. We hope you and your organization will feel free to adapt and improve upon what we have presented in the spirit of PDSA. Some of the principles are nonnegotiable if you want to call your approach Kaizen, but there is room for variation and creativity in some of the specific techniques and methods.

Now that you have read the book, it is time to take action (if you have not already done so). Regardless of your role or level, one of the first things you can do is to find a problem or opportunity within the scope of your own work. Develop an idea, and talk with your coworkers and other leaders. Ask for their input, and work to find something to implement, measure or gauge the impact, and consider whether the change is really an improvement. Remember to follow the scientific method and the PDSA process. Document your improvement, and share it with others. With your participation and leadership, you will set off a wave of Kaizen throughout your organization, leading to broader program and a thriving Kaizen culture.

Your ongoing leadership and direct involvement is necessary to grow and sustain this culture. It takes more than one all-hands meeting speech or one health system newsletter pronouncement. As Kaplan, Toussaint, and other senior leaders have emphasized, this requires looking in the mirror and changing behaviors. What past behaviors have stifled a Kaizen culture? What new behaviors will continue to add energy to the system? We hope this book has showed you the way.

Building a Kaizen Community

We wrote about sharing within your organization—your department, your clinic, your hospital, or your health system. We hope that you will consider being a part of a broader Kaizen community. Does your state, province, or country have venues for sharing and collaboration? If not, can you help create them?

Please participate and encourage your staff and leaders to join in the Kaizen community we are building at our website, http://www.HCkaizen.com. We hope you will share your methods, your lessons learned, and even your Kaizens. Would our broader healthcare system improve more quickly if we did more to share Kaizens across organizational boundaries? Asking that question is part of our own personal PDSA cycle, as authors. We hope that you will join us.

Best wishes to you and your organization in using Kaizen to further your important mission.

Endnotes

1 Gawande, Atul, *Better: A Surgeon's Notes on Performance* (New York: Picador Publishing, 2008), 246.
2 Virginia Mason Medical Center, "Using Lean Ideas in Our Everyday Work," http://virginiamasonblog.org/2013/01/30/using-lean-ideas-in-our-everyday-work/ (accessed January 31, 2013).
3 Virginia Mason Medical Center.
4 Grove, Andrew S., *Only the Paranoid Survive* (London: Profile Books, 1998), 152.
5 Liker, Jeffrey, K., and David P. Meier, *Toyota Talent* (New York: McGraw-Hill, 2007), 3.
6 Brody, Bob, personal interview, January 2013.

Index

H

I

J

K

W

Y